MECHANICAL ENGINEERING THEORY AND APPLICATIONS

ADVANCES IN AEROSPACE SCIENCE AND TECHNOLOGY

PART II

MECHANICAL ENGINEERING THEORY AND APPLICATIONS

Additional books and e-books in this series can be found on Nova's website under the Series tab.

MECHANICAL ENGINEERING THEORY AND APPLICATIONS

ADVANCES IN AEROSPACE SCIENCE AND TECHNOLOGY

PART II

PARVATHY RAJENDRAN
AND
MOHD ZULKIFLY ABDULLAH
EDITORS

Copyright © 2019 by Nova Science Publishers, Inc.

All rights reserved. No part of this book may be reproduced, stored in a retrieval system or transmitted in any form or by any means: electronic, electrostatic, magnetic, tape, mechanical photocopying, recording or otherwise without the written permission of the Publisher.

We have partnered with Copyright Clearance Center to make it easy for you to obtain permissions to reuse content from this publication. Simply navigate to this publication's page on Nova's website and locate the "Get Permission" button below the title description. This button is linked directly to the title's permission page on copyright.com. Alternatively, you can visit copyright.com and search by title, ISBN, or ISSN.

For further questions about using the service on copyright.com, please contact:
Copyright Clearance Center
Phone: +1-(978) 750-8400   Fax: +1-(978) 750-4470   E-mail: info@copyright.com.

NOTICE TO THE READER

The Publisher has taken reasonable care in the preparation of this book, but makes no expressed or implied warranty of any kind and assumes no responsibility for any errors or omissions. No liability is assumed for incidental or consequential damages in connection with or arising out of information contained in this book. The Publisher shall not be liable for any special, consequential, or exemplary damages resulting, in whole or in part, from the readers' use of, or reliance upon, this material. Any parts of this book based on government reports are so indicated and copyright is claimed for those parts to the extent applicable to compilations of such works.

Independent verification should be sought for any data, advice or recommendations contained in this book. In addition, no responsibility is assumed by the Publisher for any injury and/or damage to persons or property arising from any methods, products, instructions, ideas or otherwise contained in this publication.

This publication is designed to provide accurate and authoritative information with regard to the subject matter covered herein. It is sold with the clear understanding that the Publisher is not engaged in rendering legal or any other professional services. If legal or any other expert assistance is required, the services of a competent person should be sought. FROM A DECLARATION OF PARTICIPANTS JOINTLY ADOPTED BY A COMMITTEE OF THE AMERICAN BAR ASSOCIATION AND A COMMITTEE OF PUBLISHERS.

Additional color graphics may be available in the e-book version of this book.

Library of Congress Cataloging-in-Publication Data

ISBN: 978-1-53615-689-8

Published by Nova Science Publishers, Inc. † New York

CONTENTS

Preface		vii
Chapter 1	Application of High-Performance Interconnection in Aerospace Technology *M. S. Abdul Aziz, M. H. H. Ishak and Mohd Zulkifly Abdullah*	1
Chapter 2	Knitted Structures in Aerospace Applications *Habib Awais and Mohd Shukur Zainol Abidin*	35
Chapter 3	Carbon Nanotube-Reinforced Hierarchical Carbon Fibre Composites *Mohd Shukur Zainol Abidin*	51
Chapter 4	Influence of Aviation Fuel on Composite Materials *Sharmendran Kumarasamy, Nurul Musfirah Mazlan and Aslina Anjang Ab Rahman*	73
Chapter 5	Deterioration in Aero-Engines *Koh Wei Chong, Senebahvan Maniam, Aslina Anjang Ab Rahman and Nurul Musfirah Mazlan*	123

Chapter 6	Important Aerodynamic Parameters of Flapping-Wing Unmanned Aerial Vehicles *N. A. Razak, Aizat Abas and Zarina Itam*	**149**
Chapter 7	Visual Localisation and Mapping Using Unmanned Aerial Vehicles *Kai Yit Kok, Parvathy Rajendran, Nurulasikin Mohd Suhadis and Muhammad Fadly*	**167**
Chapter 8	Geospatial Mapping Using Satellites *Mohd Badrul Salleh, Nurulasikin Mohd Suhadis, Parvathy Rajendran and Hassan Ali*	**185**
About the Editors		**215**
Index		**219**
Related Nova Publications		**225**

PREFACE

This book provides a quick read for experts, researchers, and novices in the field of aerospace research, science, technology, and applications, theory, and trends in research. The relevant subject areas include mechanical engineering, material sciences, computational engineering, and system design. This book includes the study of theoretical and practical groundwork as well as techniques, methods, and processes currently used in aerospace research. This book is intended for university undergraduate and postgraduate students.

The topics include aeroelasticity, nanocomposites and fire structural behavior of aerospace composites, implementation of biofuel in aircraft engines, aircraft mission profile and trajectory, hybrid UAV controller design and magnetic attitude control of small satellites, CubeSat technology, ionosphere perturbation modeling by electromagnetic wave propagation, and simulation of membrane airfoil with fluid/structure interaction.

A group of lecturers from the School of Aerospace Engineering and School of Mechanical Engineering at Universiti Sains Malaysia collectively wrote the book in a simple and easy-to-understand language for engineering students. Our years of experience in teaching and research on these subjects will surely benefit and assist students. We hope that this book will communicate to students our experiences and inspire enthusiasm among students in engineering disciplines.

The authors would like to thank all their colleagues in Malaysia and overseas who have contributed to this book for their support and comments. The book is intended for practical engineers, researchers, students, and people who would like to be enlightened about the recent advancements in aerospace science and technology.

In: Advances in Aerospace Science ...   ISBN: 978-1-53615-689-8
Editors: Parvathy Rajendran et al.   © 2019 Nova Science Publishers, Inc.

Chapter 1

APPLICATION OF HIGH-PERFORMANCE INTERCONNECTION IN AEROSPACE TECHNOLOGY

M. S. Abdul Aziz[1,], M. H. H. Ishak[2] and Mohd Zulkifly Abdullah[1]*

[1]School of Mechanical Engineering, Universiti Sains Malaysia, Pulau Pinang, Malaysia
[2]School of Aerospace Engineering, Universiti Sains Malaysia, Pulau Pinang, Malaysia

ABSTRACT

Wave soldering is one of the most familiar and well-established processes in the electronics assembly industries, and it is used to assemble pin-through hole (PTH) components onto printed circuit boards (PCBs). The reliability of a solder joint is essential in ensuring the functionality of the component as it performs intended functions in electrical, mechanical and thermal conditions. Solder joint failure during microelectronics assembly is a challenge and a major concern of engineers in maintaining product reliability. Solder joint defects, such as cracks, void formation and

---

[*] Corresponding Author's Email: msharizal@usm.my.

incomplete filling of the PCB hole, may weaken the PTH solder joint. Improper control of process and physical parameters may induce unintended issues in PTH. In this research, PTH vertical fill was applied experimentally using a newly developed adjustable fountain wave soldering machine (0° conveyor angle). Similar PCB and components were assembled using a conventional wave soldering machine (6° conveyor angle). A non-destructive X-ray computed tomography–scanning machine was employed to inspect the vertical fill of each solder joint. Results showed that the adjustable fountain wave soldering process yielded a higher vertical fill (~ 99.4%) at a 0° conveyor angle than the conventional wave soldering process (~ 90.8%) at a 6° conveyor angle. Simulation modelling facilitates predictions of real situations in small-scale and complex geometry. Therefore, this research focused on the wave soldering process considering the thermal fluid–structure interaction (FSI) phenomenon. A thermal FSI simulation was performed using finite volume- (FLUENT) and finite element (ABAQUS)-based software through a real-time coupling technique using Mesh-based Parallel Code Coupling Interface software. The volume of the fluid model tracked the flow fronts of the molten solder, such as filling profile, wetting pattern and wetting area, in the fountain flow analysis. The solder pot temperature of 523 K showed good performance and achieved 90% of the full wetting area. The simulation was broadened to parametric studies on various processes and design factors, such as conveyor angle, PTH geometry (i.e., offset position, shape, diameter and offset angle), soldering temperature and PCB thickness. The effects of these parameters on fluid flow distribution, void formation, structural displacement, pressure distribution and stress were highlighted. The correlation among the parametric studies was found to influence PTH and PCB temperature, displacement and stress. Moreover, optimisation of the PTH connector in the wave soldering process using response surface methodology was conducted to study the interactive relationship between independent variables and responses (i.e., filling time at 75% volume, von Mises stress and maximum displacement). The optimum geometrical and process parameters for PCB and PTH connectors were characterised by 0.12 mm PTH offset position, 0.17 mm PTH diameter, 0° offset angle and 473 K molten solder temperature. A case study was conducted on the effects of PCB configuration during wave soldering. Five PCB configurations were considered based on the position of the components. The thermal-induced displacement and stress on the PCB and its components were the foci of this study. The results revealed that PCB component configurations significantly influenced the PCB and yielded unfavourable deformation and stress. The current research findings are expected to provide valuable guidelines and significant contributions to the wave soldering process for the microelectronics industry.

Keywords: wave soldering, solder joint, aerospace

1. INTRODUCTION

In the past several decades, the solders used for the assembly of high-reliability electronics were predominantly made of an alloy of tin (Sn) and lead (Pb). The performance characteristics of this alloy and the associated manufacturing/repair processes are well understood. Reliability analyses and qualification tests have been developed; these can adequately predict and verify system performance to ensure that the required system reliability is achieved. The rapid development of microelectronic technologies presents additional challenges in ensuring the reliability and quality of electronic assemblies. The high demand for miniature, lightweight and high-performance electronic products directs the focus towards the reliability of interconnectors between electronic components and printed circuit boards (PCBs). High solder joint reliability provides superior mechanical bonding and component functionality. The mechanical strength of the pin-through-hole (PTH) component is the reason for implementing through-hole technology (THT) in assembling industries (Berntson, Lasky, & Pfluke, 2002). The PTH component is assembled in the wave soldering process, which necessitates the use of a particular wave soldering machine to solder PTH onto the PCB.

The ideal assembly process in wave soldering achieves an optimum solder joint result without damaging its parts or the assembly in any way and presents almost zero-defect conditions. Definitive design efforts and proper process control of parameters, including standardised solder joints, minimum solder spikes or surplus solder, solder skips, bridges and minimum cleanliness of the assemblies, are required to attain this goal (Bergenthal, 2014; Martin, 2014). Thus, an optimum solder fillet and zero repairs can be achieved. The PCB and component damage due to the over stress and cleaning of the product can also be minimised. The responsibility of the circuit designer to achieve a zero-defect PCB and solder joint in the assembly process becomes challenging due to the increase in component densities, board thickness and fine-pitch devices. An acceptable PTH solder fillet with practical yields, costs and process improvement is the key to a successful board design (Berntson et al., 2002). Changes in surface finish, increasing power request, and reliability issues are also important. Thus, strict requirements on electronic packaging are required in PCB assemblies.

Electronic equipment in the aerospace industry encounters unique challenges, such as rugged operating environments and high likelihood of failure. According to the Institute for Printed Circuits (IPC-A-610E, 2010), the acceptance and rejection of solder joints are decided by referring to applicable documentation, such as standards, drawings, contracts and references. The criteria for reference documents are classified into three classes (1 to 3). For class 1, the required criteria are applicable to general electronic products. Class 2 is specified for dedicated service electronic products, and class 3 is for high-performance electronic products (IPC-A-610E, 2010). These circuit boards are highly reliable and used to achieve high performance in military, aerospace or medical areas. For class 1 and 2 boards, the minimum acceptable solder joint must present 270° of circumferential fillet (IPC-A-610E, 2010); the minimum acceptable solder joint for class 3 boards is 330° (Alpha, 2011). Figure 1 illustrates a solder joint circumferential fillet and wetting for targeted, acceptable and defective PTH solder joints after wave soldering. Examples of poor wetting and lead solderability as wave soldering defects are shown in Figure 2 (Epec, 2014).

1.1. PCB Assembly

Soldering is the most important process in the microelectronics industry. As the demand for electronic products increases, soldering is transformed from the conventional method to machine soldering to improve quality, reliability and process rate. Reflow and wave soldering are the two main soldering processes involved in surface mount technology (SMT) and THT.

Reflow soldering is the process of attaching surface mount components (SMCs) to PCB (Figure 3(a)). The preparatory steps involve screen printing of solder paste to the PCB bond pads, followed by the placement of SMCs on the solder paste deposit. The PCB assembly is then subjected to controlled heat in a reflow oven, which melts the solder paste and solder balls and permanently forms the joint (Koch, 1998). The controlled heat in a reflow oven is programmed according to the reflow thermal profile. A reflow thermal profile has eight output elements, namely, preheating slope, soaking temperature, ramp-up slope, peak temperature and durations of the four heating periods (C. Lau, Abdullah, & Che Ani, 2012; Tsai, 2009). The drawback of using SMT is the failure of the solder joint in handling high-power devices, especially in military and aerospace industries. These applications require additional strength for the solder joints (Backwell, 2006).

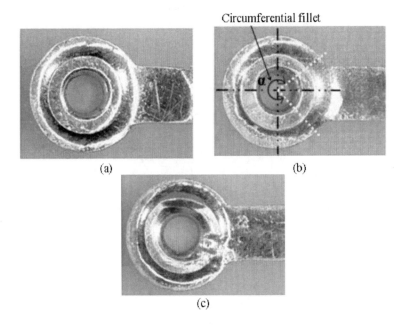

Figure 1. Solder joint circumferential fillet and wetting: (a) targeted, (b) acceptable and (c) defective (IPC-A-610E, 2010).

THT is implemented to overcome these problems. The method of attaching PTH onto PCB is wave soldering, which is a large-scale soldering process used in through-hole component (THC) assemblies. Typically, PCBs with inserted PTHs are pre-fluxed, preheated and passed over a dual solder wave for soldering (N.-C. Lee, 2002). A soldered PTH on PCB is illustrated in Figure 3(b). The details of the wave soldering process are discussed in the next subchapter.

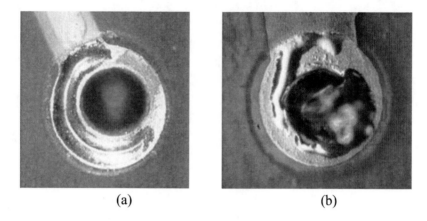

Figure 2. Wave soldering defects: (a) poor wetting on the inner and outer edges and (b) poor solderability on the pad surface (Epec, 2014).

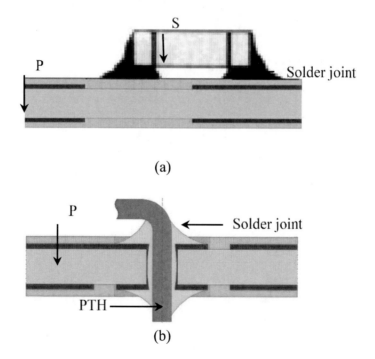

Figure 3. Types of solder joint: (a) SMT and (b) THT.

1.2. Wave Soldering

Wave soldering, a large-scale process implemented in THC assemblies, is one of the most widely used soldering methods in the electronics industry. A wave soldering machine is used to assemble the PTH components (e.g., capacitor, transistor and passive components) of a PCB. This process involves subsequent stages, such as (i) pre-fluxing the PCB with electronic components via flux spray or foam fluxes, (ii) preheating, (iii) passing through a single or dual wave for soldering and (iv) cooling (N.-C. Lee, 2002). In the wave soldering zone, a PCB with PTH components experiences high-temperature molten soldering. The bottom surface of the PCB with PTH components interacts with the fountain of the molten solder whilst the molten solder fills the PCB hole and is driven by the capillary force of the PCB hole. Afterwards, the filled PCB hole is solidified in the cooling zone. Understanding the process parameters of wave soldering, such as temperature, conveyor speed, soldering angle, preheating and cooling zones, is imperative in achieving optimal process conditions. However, different designs of wave soldering machines (e.g., for low- to high-volume production and various zones) may have different optimum process parameter controls. A schematic of the conventional wave soldering process is shown in Figure 4.

In most industrial applications, a dual wave soldering machine is used to minimise solder joint defects, such as incomplete filling, solder bridge, voids and non-wetting of lead. The configuration of dual-wave soldering is illustrated in Figure 5. The PCB undergoes a primary or 'chip' wave with a high-energy turbulent peak to ensure that the molten solder meets every joint of the PTH (Martin, 2014). Next, the PCB passes through a secondary or 'lambda' wave, which removes solder bridges or accumulated solder by removing the excess solder from the PCB. The mechanisms of the chip and lambda wave are shown in Figures 6 and 7, respectively.

The wave soldering process has been widely utilised in the high-volume soldering of PCBs in electronic assemblies (Bertiger & Mesa, 1985). The process involves fluxing, preheating, soldering and cooling. This process is used to solder and assemble PTH components (e.g., capacitor, resistor, transistor and pin connector) onto PCBs. However, components without a pin, such as the ball grid array (BGA) integrated circuit (IC) package and leadless components, are mounted onto PCBs through a reflow soldering process (LoVasco & Oien, 1983). Typically, reflow soldering is performed to solder the leadless component (e.g., surface mount capacitor and resistor). The process is followed by PTH component placement and wave soldering.

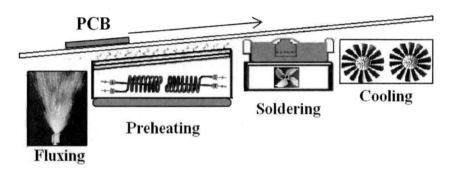

Figure 4. Conventional wave soldering process (N.-C. Lee, 2002).

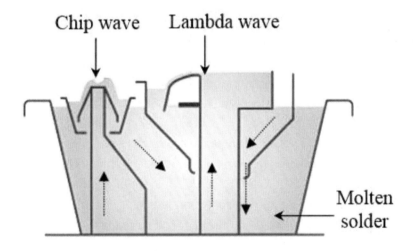

Figure 5. Dual wave soldering process (Martin, 2014).

Application of High-Performance Interconnection in Aerospace ...

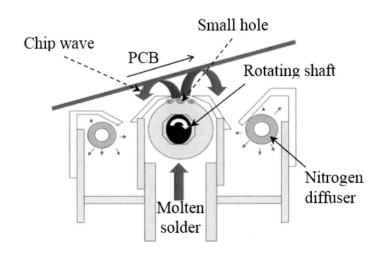

Figure 6. Chip wave.

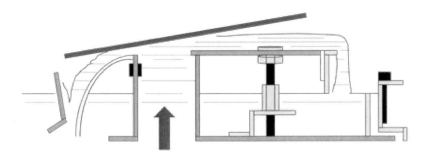

Figure 7. Lambda wave.

During this process, the earlier mounted component and the PCB undergo second thermal loading (Franken, Maier, Prume, & Waser, 2000) from the high-temperature environment. Improper temperature control in the wave soldering process can induce unintended deformation and stress in the PCB and leadless component. Therefore, understanding the thermal profile and phenomenon during the wave soldering process is essential in minimising and eliminating the defects or unintended features of PCBs and their components. Figures 8 and 9 depict the experimental setups of the wave soldering process utilised by (M. Liukkonen, Havia, Leinonen, & Hiltunen, 2009; Mika Liukkonen, Havia, Leinonen, & Hiltunen, 2011a) and (D. W. Coit, Jackson, & Smith, 1998) in their investigations.

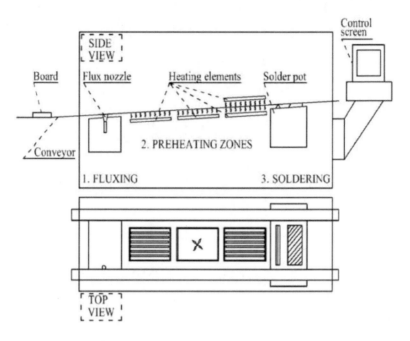

Figure 8. Schematic of wave soldering (M. Liukkonen et al., 2009; Mika Liukkonen et al., 2011a).

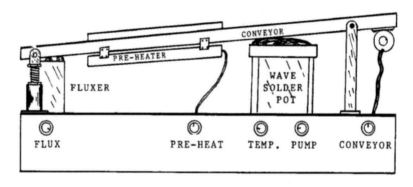

Figure 9. Wave soldering process (D. W. Coit et al., 1998).

The majority of wave soldering studies were experimental investigations. These studies reported on the characteristics of solder materials. Several other studies examined mechanical characteristics (Baylakoglu, Hamarat, Gokmen, & Meric, 2005; Kuo et al., 2013), low-silver alloy (J. Wang, Wei, Zhu, Wu, & Wu, 2013), self-organising maps

(SOMs) of process optimisation (M. Liukkonen et al., 2009; Mika Liukkonen, Havia, Leinonen, & Hiltunen, 2011b), design of experiment (DOE) for optimisation of materials and process parameters (Arra, Shangguan, Yi, Thalhammer, & Fockenberger, 2002; D. W. Coit et al., 1998; Huang & Huang, 2012), thermal expansion (Fu & Ume 1995), lift-off phenomenon (Suganuma et al. 2000), stress induced by thermal shock (Franken et al. 2000), soldering contact time (Morris & Szymanowski, 2008), design of the PTH hole (S. Chang, Wang, Xiang, Wang, & Shi, 2011) and gold concentration of the pin (Che Ani, Abdul Aziz, Jalar, Abdullah, & Rethinasamy, 2012).

In the wave soldering process, the filling of the PCB hole is also known as vertical fill (IPC-A-610E, 2010), which is commonly used in the electronics assembly industry. The percentage of vertical fill crucially influences the strength of the solder joint. Therefore, maximum vertical fill (100%) is favoured in the wave soldering process, as shown in Figure 10(a). The minimum vertical fill of 75% (Figure 10(b)) is acceptable; otherwise, the PTH solder joint is considered defective. The vertical fill requirement is strict in applications (e.g., aerospace and military applications) that involve high thermal shock and electrical performance. The solder joint is considered defective when the vertical fill is less than 100% in these applications. However, a vertical fill of 50% (Figure 10(c)) is acceptable under specific conditions when PTH is connected to thermal or conductor layers that function as thermal heat sinks (IPC-A-610E, 2010). Other problems, such as bridging, insufficient topside and bottom side fillet, may lead to short circuit and weak solder joints. Therefore, proper control of process parameters (Alpha, 2013) (e.g., solder wave height, soldering dwell time and preheating time) is crucial in eliminating these features of the PTH solder joint.

Solder joint failure typically occurs in the interconnection between a component (e.g., IC package and passive component) and mounting boards (e.g., solder bumps, solder studs and PTH connectors). The thermal and metallurgical reliability of this solder joint is one of the major issues that hinder the development of small, high-density interconnections (Shi, Zhou, Pang, & Wang, 2014). The stresses and displacement caused by PTH on solder joints may result in solder joint defects. Improper control of molten solder temperature leads to solder joint fractures when mounting onto PCB.

This situation subjects the PTH to thermo–mechanical stress that exceeds the fracture strength of the solder joint. The density and viscosity of the molten solder also significantly influence the reliability of wave soldering (Abtew & Selvaduray, 2000). Moreover, the molten solder filling characteristics may influence the quality of the solder joint (e.g., unbalanced flow front may induce incomplete filling). Therefore, understanding the molten solder flow characteristics and process visualisation is crucial for engineers to sustain solder joint reliability.

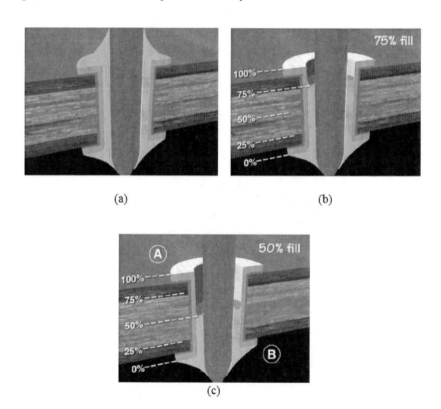

Figure 10. Vertical fill of solder joint (IPC-A-610E, 2010): (a) 100% fill, (b) minimum 75% fill and (c) 50% fill.

2. SIMULATION MODELLING

Simulation analysis plays a vital role in engineering applications, such as in the modelling of biomechanical devices, aircraft structures, automotive components and microelectronic devices. The rapid development of virtual modelling tools in recent years has enhanced the reliability of electronic devices. With the aid of simulation modelling techniques, realistic predictions can be achieved through various modelling tools. Modelling tools, such as FORTRAN (Abdullah, Abdullah, Kamarudin, & Ariff, 2007; Abdullah, Abdullah, Mujeebu, Ariff, & Ahmad, 2010), FLUENT (C. Y. Khor, Abdullah, & Che Ani, 2011; Chu Yee Khor, Abdullah, & Leong, 2012), C-MOLD (Bae et al., 2003; R.-Y. Chang, Yang, Hwang, & F, 2004), PLICE-CAD (Hon, Lee, Zhang, & Wong, 2005), Moldex3D (H. Wang, Zhou, Zhang, & Li, 2010), computer-aided engineering and finite element (FE)-based software (Jong, Chen, & Kuo, 2005; Pei & Hwang, 2005; Teng & Hwang, 2008), have been employed in microelectronics research.

These modelling tools were mainly used for fluid flow predictions and structural analyses. Optimal design, process setting and material selection of the wave soldering process can be achieved before mass production. Complex governing equations integrated with the volume of fluid (VOF) technique for flow front tracking have been solved by using the finite difference method (Hashimoto, Shin-ichiro, Morinishi, & Satofuka, 2008), finite volume method (Shen, Huang, Chen, & Yu, 2006; Wan, Zhang, & Bergstrom, 2009) and characteristic-based split method in conjunction with the FE method (FEM) (Kulkarni, Seetharamu, Azid, Aswataha Narayana, & Abdul Quadir, 2006). Modelling tools can also solve complex problems and highly nonlinear problems that involve thermal and mass transfer, which are combined with deformation, pressure and stress on the fluid structure (Kuntz & Menter, 2004).

Experimental work is important in investigating actual problems. An experiment is sometimes limited to specific situations, such as the visualisation of deformation, temperature and stress distribution on the PCB and its components during the process. Tiny PTH and limited equipment have created difficulties in visualising the real-time PTH filling process.

Thus, simulation tools are useful in modelling and simulating the actual wave soldering process. These tools can also provide predictive trends of reliability problems. In addition, they are beneficial for small-scale and complex geometry and can minimise the cost of long-term research activities.

Recently, real-time fluid–structure interaction (FSI) and thermal-coupling methods have also been employed to solve wire deformation (Ramdan et al., 2011), reflow soldering (C. Lau et al., 2012), moulded underfill (C. Y. Khor et al., 2011; Chu Yee Khor et al., 2012) and flexible PCB (Leong, Abdullah, & Khor, 2012). Researchers used the Mesh-based Parallel Code Coupling Interface (MpCCI) coupling method to integrate FLUENT and ABAQUS. Simulations were conducted simultaneously, and real-time data transfer was handled by MpCCI software. The predictions of this coupling method yielded reliable simulation results. These studies focused on the FSI phenomenon of IC packaging (Ramdan et al., 2011; Khor et al., 2011, 2012) and the feasibility of flexible PCB for the computer motherboard (Leong et al., 2012) without considering the thermal aspect. Lau et al. (2012a, 2012b) investigated the assembly of BGA IC package during the reflow soldering process. In their work, the thermal stress and deformation of the BGA IC package were analysed by considering various factors (e.g., BGA array, shape, board level and package level). Although the FSI issue has been explored in various electronic packaging and assembly processes, a wide research gap remains in the wave soldering process. The detailed thermal–FSI phenomenon of the electronic assembly process can be revealed by using a simulation–modelling technique, which contributes to the understanding and clear visualisation of the process. The applications of real-time FSI simulation using MpCCI coupling software by researchers are presented in Figures 11 to 13.

Application of High-Performance Interconnection in Aerospace ... 15

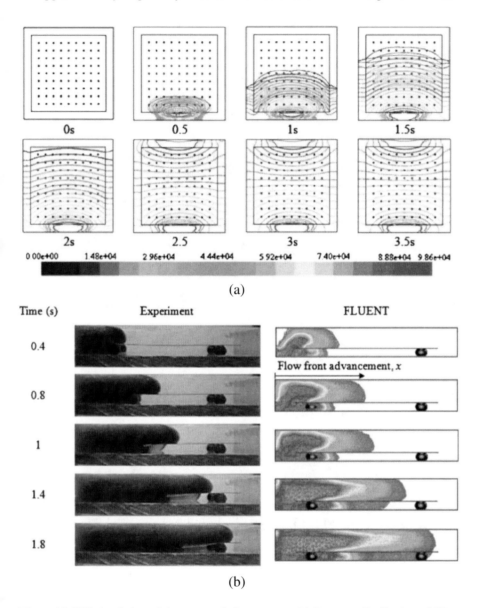

Figure 11. FSI simulation of the encapsulation process. (a) Pressure distribution of 3D moulded package at 0.1 MPa inlet pressure (C. Y. Khor et al., 2011). (b) Comparison of scaled-up experimental and simulation results of flow front advancement in moulded flip-chip package (C. Y. Khor, Abdullah, Ariff, & Leong, 2012).

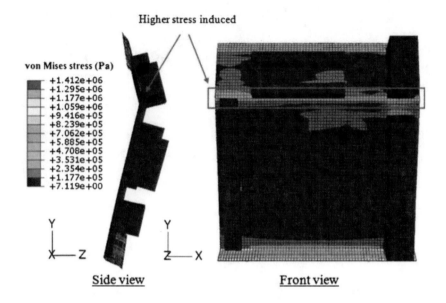

Figure 12. FSI investigation of flexible PCB in computer motherboards (Leong et al., 2012). Stress distribution contours (flow velocity = 3 m/s).

2.1. Solder Material

In wave soldering, many factors, such as the physical design of PCB, process control (i.e., solder temperature and conveyor speed) and materials used, could influence the process performance and flowability of the molten solder in the PCB hole. The majority of wave soldering studies were based on experimentation; they usually focused on the solder joint and materials used. The characteristics of solder materials and the filling of the molten solder in PCB holes may influence the reliability and quality of solder joints.

Lead-free (Sn/Ag/Cu) wave soldering that enables environment-friendly and high-volume production was proposed by (Baylakoglu et al., 2005), who investigated the significance of nitrogen in dross rate reduction, product quality and process robustness. The experimental setups of wave soldering equipment used by (Baylakoglu et al., 2005) are shown in Figure 14. Dross formation (Sn/Ag/Cu) in a nitrogen environment was compared with a eutectic leaded solder (Sn/Pb) bath under an air atmosphere. Board quality was also evaluated via the application of no-clean, volatile organic compound-free flux. The researchers found that the application of nitrogen with Sn/Ag/Cu alloy enhances the wave soldering process. Increasing the solid quantity of flux yields improved solder joints and minimises short rates. Although a lead-free solder is environment-friendly, time is required for the transition from leaded to lead-free solder materials in the industry.

An experimental investigation of lead-free wave soldering using 95.5Sn/3.8Ag/0.7Cu alloy was conducted by (Arra et al., 2002), who performed DOE in consideration of three variables, namely, solder bath temperature, conveyor speed and soldering atmosphere, in a dual-wave system (Figure 15) and four no-clean flux systems, including alcohol- and water-based types. A 'lead-free solder test vehicle' was used in these experiments.

The water-based flux system was found to be the best fluxing system for various process conditions. The soldering results indicated that optimum flux and process parameters are more important than Sn/Pb in achieving acceptable Sn/Ag/Cu soldering performance. No fillet lifting was observed in any tested components and PCB surface finishes. The cross-sectional view of several components coated with Sn/Pb and soldered with Sn/Ag/Cu and Sn/Cu is provided in Figure 15.

3. PCB WARPAGE AND SOLDER JOINT DEFECTS

PCBs generally undergo different thermal expansions when they pass through wave soldering; this phenomenon leads to thermomechanical fatigue, manufacturing defects and mechanical catastrophic failure of solder joints (Shen et al., 2006). Warpage occurs when the PCB temperature exceeds the temperature of glass transition (Fu & Ume, 1995), and it leads to other assembly problems, such as non-coplanarity, solder joint failure, floating, poor connectivity and chip/component surface cracking (Polsky, Sutherlin, & Ume, 2000).

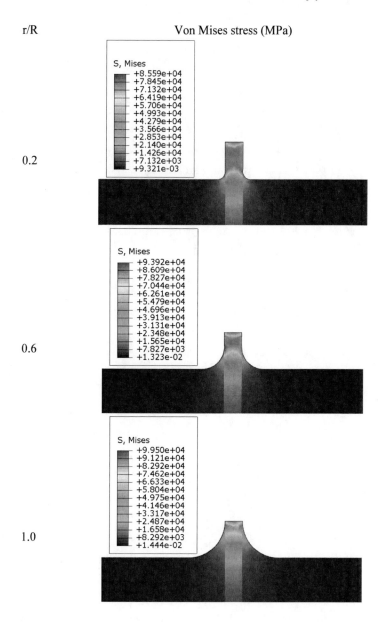

Figure 13. Effect of fillet on the stress of a solder joint (0.2 < r/R < 1.0).

PCB plays an important role in electronic devices and components. PCBs are fabricated from different organic materials that are printed with circuits, such as polyimide, FR-1, FR-2, FR-4, bismaleimide and cyanate ester. Complex PCBs can be classified into single-sided, double-sided and multi-layered. Printed circuits provide power and connect the components and IC package. PCBs function as carriers for components and IC packages. In the assembly process, a PCB experiences temperature variation in different assembly stages, such as reflow and wave soldering. The temperature at reflow and wave soldering is approximately 240°C to 250°C. During these processes, the PCB temperature increases and decreases in each stage. These temperature variations may induce PCB warpage, which is also recognised as out-of-plane displacement (Tan & Ume, 2012). The coefficient of thermal expansion mismatch among the materials results in PCB warpage. PCB warpage is a concern not only for board transfer during automated assembly but also for the performance of the final assembly because it may lead to severe joint failures.

Multifarious studies have been conducted to investigate PCB warpage. However, most of these studies concentrated on the solder reflow process through experiments (Cepeda-Rizo, Yeh, & Teneketges, 2005) and simulations (Ume, Martin, & Gatro, 1997). Many scholars used FEM (Hutapea, Grenestedt, Modi, Mello, & Frutschy, 2006; M. Lee, 2000; Ume et al., 1997) to investigate PCB and substrate warpage. However, an investigation of PCB deformation using thermal FSI is yet to be conducted. Warpage investigations have been extended to IC packages (Driel et al., 2003), such as flip-chip BGA packages. Reference (Ding, Ume, Powell, & Hanna, 2005) studied the parametric effects of the materials, geometry and process of PCB assembly during reflow. The Taguchi method and full-factorial DOE were used to optimise parameters and minimise warpage. Shadow Moiré (Ying, Chia, Mohtar, Yin, & Chuah, 2006) is a popular method used in experimental work to measure PCB warpage. The deflection of PCB is evaluated through the grating captured by the system. However, in simulation modelling analysis, most of the warpage and solder joint (Wong & Wong, 2009) are predicted through FEM, which only involves a structural solver. Temperature endurance to PCB depends on the input setting.

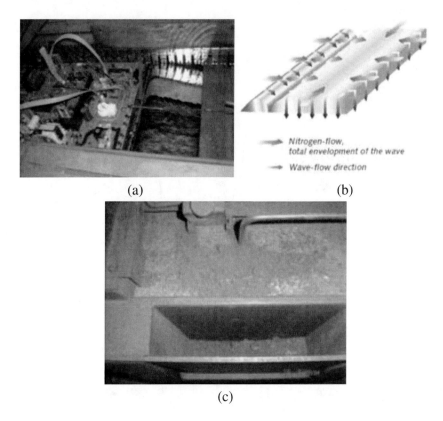

Figure 14. Experimental setup of (Baylakoglu et al., 2005). (a) Wave soldering machine. (b) Nitrogen and N_2 flow direction. (c) Dross-removing pot.

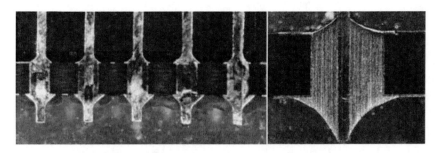

Figure 15. Cross-sectional view of pin connectors (Arra et al., 2002).

Only a few scholars have studied PCB warpage and solder joint defect during wave soldering. Reference (Polsky et al., 2000) investigated PCB warpage caused by infrared reflow and wave soldering. Two PCB layouts were used to study thermally induced deflection. They observed that both PCBs experienced large deflection (bowl-like shape) in wave soldering due to the presence of large through-thickness thermal gradients. PCB deflection was evaluated using a shadow Moire pattern, as illustrated in Figure 16. To minimise warpage, (Johnson, 2014) reported the advantages of a pallet during wave soldering. They introduced a high-temperature thermoset composite as a solder pallet material, as depicted in Figure 17.

The pallet exposes a particular PCB region (soldering area) to the solder wave, which can minimise PCB warpage during wave soldering. When the pallet absorbs heat from the solder wave, only a portion of the heat is absorbed by PCB. This method can minimise PCB warpage, but the pallet is warped or twisted when it is exposed to high temperature. The pallet design and material used are important in solving this problem. Moreover, (Hutapea & Grenestedt, 2004) reported that the wavy copper trace design (Figure 18) of PCBs can minimise 40% to 60% of warpage compared with a typical straight copper trace. The experimental and FE analyses of PCB warpage versus time by (Hutapea & Grenestedt, 2004) are shown in Figure 19.

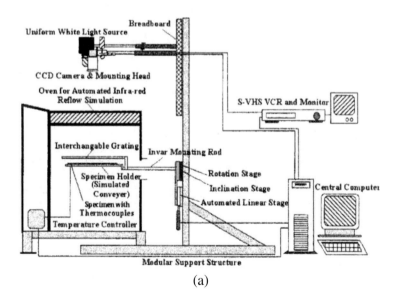

(a)

Figure 16. (Continued).

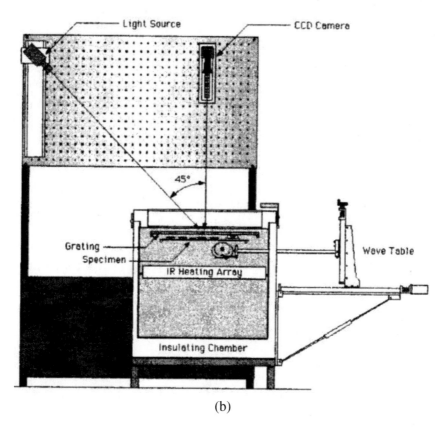

(b)

Figure 16. Shadow Moire online measurement setup implemented by (Polsky et al., 2000). (a) Infrared reflow soldering. (b) Wave soldering.

Figure 17. Solder pallet material and pallet design suggested by (Johnson, 2014). (a) High-temperature thermoset composite material. (b) Lead-free wave soldering pallet.

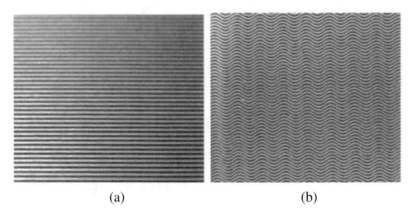

(a) (b)

Figure 18. PCBs used in the work of (Hutapea & Grenestedt, 2004). (a) Straight PCB. (b) Wavy PCB.

The parametric design of PCBs, including mass, size, component density and component type, influences the control over the wave soldering process (David W Coit & Smith, 1995). Different component configurations affect the temperature distribution and thermal profile of PCBs. Visualising the PCB temperature contour, heat transfer coefficient and stress during the process is difficult because of limited equipment. The coupling of different solvers yields an effective model of the process phenomenon through the exchange of temperature and heat transfer coefficient data from the thermal fluid and structural analyses (C.-S. Lau, Abdullah, Ani, & Che Ani, 2012).

As mentioned previously, PCB warpage induces unintended defects on the component and solder joint and may cause the malfunction of electronic devices. Therefore, minimisation of PCB warpage is crucial in eliminating solder joint defects. Scholars have investigated several solder joint defects. One of the defects is a lift-off mechanism, which is a type of severe defect formation as shown in Figure 20, studied numerically by (Suganuma et al., 2000). This problem occurs when the solder fillet peels off from a Cu land pad on a PWB. In this study, basic Sn–Bi alloys (Sn at 2 wt% to 5 wt%) and the solidification simulation method were used. The researchers found that the lift-off problem results from the rapid solidification of the solder fillet, which propagates from a lead wire to a fillet edge. This problem was mitigated through rapid cooling and annealing at a high temperature.

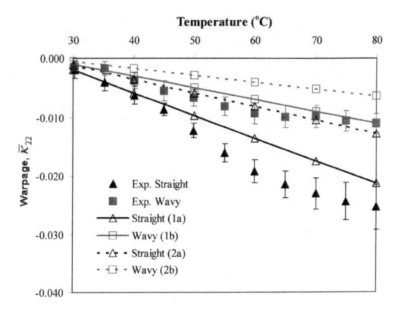

Figure 19. PCB warpage against temperature (°C) in the experiment and predicted FE results of (Hutapea & Grenestedt, 2004).

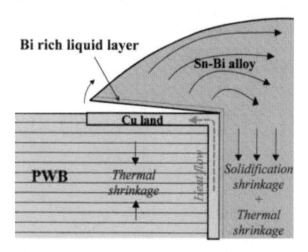

Figure 20. Lift-off mechanism in wave soldering (Suganuma et al., 2000).

Application of High-Performance Interconnection in Aerospace ... 25

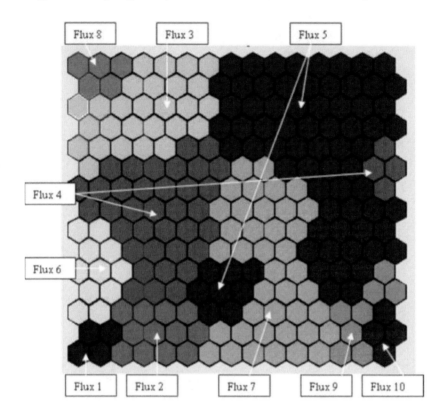

Figure 21. SOM using wave soldering (M. Liukkonen et al., 2009).

In the wave soldering process, the PCB experiences different thermal expansions (CTEs). This situation causes manufacturing defects and fatigue in the PCB during field operations (Guo, 1995). Adding anti-oxidation elements in solder alloys improves oxidation resistance but results in poor wettability in low-silver, lead-free solder (J. Wang et al., 2013). A combination of many factors in the PCB design and process requires proper process control in the wave soldering process. An intelligent system (M. Liukkonen et al., 2009; Pietraszek, Gadek-Moszczak, & Torunski, 2014) identifies defects to help engineers reduce defects in the wave soldering process, such as SOM by (M. Liukkonen et al., 2009), as presented in Figure 21. The system also identifies the dominant process parameters that cause defects during wave soldering. Hence, optimisation of the wave soldering process by considering physical and process parameters is important in enhancing the quality and reliability of PCB and solder joints.

Conclusion

The application of a circular pin for PTH components provided an almost uniform filling profile and the highest filling level compared with those of other pin shapes. The circular pin also yielded the fastest filling time. However, the triangular and hexagonal PTH pins yielded low filling levels in the filling process and created an unbalanced molten solder profile. The edge shape of a pin influenced surface tension and adhesive force, resulting in a slow flow front around the pin. The effects of molten solder temperature ranging from 456.15 K to 643.15 K were also investigated using thermal FSI modelling. The filling time exponentially decreased as the temperature increased. The temperature profile of the pin indicated that the pin reached the temperature of the molten solder during filling, which corresponded to the capillary flow. The results revealed that filling time exhibited a quadratic behaviour with the increment in temperature. Pin deformation showed a linear correlation with temperature. Furthermore, increased pin offset angle resulted in an unbalanced capillary flow profile, variations in pressure and increased filling time. High pin displacement was caused by the high thermal stress of the conduction effect over a long filling time. This study showed that a 0° offset angle is the appropriate pin position in terms of filling profile, filling time, displacement and stress.

Acknowledgments

The authors would like to extend their appreciation to Universiti Sains Malaysia (Research University Grant (RUi) 1001/PMEKANIK/8014072) for providing the financial and technical support for this research.

REFERENCES

Abdullah, M. K., Abdullah, M. Z., Kamarudin, S. & Ariff, Z. M. (2007). Study of flow visualization in stacked-Chip Scale Packages (S-CSP). *International Communications in Heat and Mass Transfer*, *34*(7), 820–828. http://doi.org/10.1016/j.icheatmasstransfer.2007.04.003.

Abdullah, M. K., Abdullah, M. Z., Mujeebu, M. A., Ariff, Z. M. & Ahmad, K. a. (2010). Three-dimensional modelling to study the effect of die-stacking shape on mould filling during encapsulation of microelectronic chips. *IEEE Transactions on Advanced Packaging*, *33*(2), 438–446. http://doi.org/10.1109/TADVP.2009.2034013.

Abtew, M. & Selvaduray, G. (2000). Lead-free Solders in Microelectronics. *Materials Science and Engineering: R: Reports*, *27*(5–6), 95–141. http://doi.org/10.1016/S0927-796X(00)00010-3.

Alpha. (2011). *Alpha Wave soldering troubleshooting guide, Easy-to-use advice for common wave soldering assembly issues*.

Alpha. (2013). *Wave soldering troubleshooting guide*.

Arra, M., Shangguan, D., Yi, S., Thalhammer, R. & Fockenberger, H. (2002). Development of Lead-Free Wave Soldering Process. *IEEE Transactions on Electronics Packaging Manufacturing*, *25*(4), 289–299.

Backwell, G. R. (2006). *Surface Mount Technology* (DORF, R.C.). Boca Raton: Taylor & Francis Group.

Bae, D. H., Lee, M. C., Lee, E. S., Yun, H. C., Lim, J. C. & Kim, I. B. (2003). Simulation of Encapsulation Process for BGA Type Semi-conducting Microchip.pdf. *Journal of Industrial and Engineering Chemistry*, *9*(2), 188–192.

Baylakoglu, I., Hamarat, S., Gokmen, H. & Meric, E. (2005). Case Study for High Volume Lead-Free Wave Soldering Process with Environmental Benefits. In *ISEE '05 Proceedings of the International Symposium on Electronics and the Environment*, (pp. 102–106).

Bergenthal, J. (2014). *Another Look - Wave Solder Process for Surface Mount Applications Complications/Contradictions/ Excellent Performance The Key to Performance is Understanding and A Robust Part*.

Berntson, R. B., Lasky, R. & Pfluke, K. P. (2002). *Through-Hole Assembly Options for Mixed Technology Boards.*

Bertiger, B. R. & Mesa, A. (1985). *Wave soldering in a reducing atmosphere*, United States Patent 4538757.

Cepeda-Rizo, J., Yeh, H. Y. & Teneketges, N. (2005). Characterization and Modeling of Printed Wiring Board Warpage and Its Effect on LGA Separable Interconnects. *Journal of Electronic Packaging, 127*(2), 178–184. http://doi.org/10.1115/1.1898339.

Chang, R. Y., Yang, W. H., Hwang, S. J. & F. S. (2004). Three-dimensional modeling of mold filling in microelectronics encapsulation process. *IEEE Transactions on Components and Packaging Technologies, 27*(1), 200–209.

Chang, S., Wang, R., Xiang, Y., Wang, P. & Shi, W. (2011). Design for manufacturability of PTH solder fill in thick board with OSP finish. *2011 International Conference on Electronic Packaging Technology and High Density Packaging*, (1), 712–719. http://doi.org/10.1109/ICEPT.2011.6066932.

Che Ani, F., Abdul Aziz, M. S., Jalar, A., Abdullah, M. Z. & Rethinasamy, P. (2012). Effect of gold concentration through a single dynamic wave soldering process. *2012 14th International Conference on Electronic Materials and Packaging (EMAP)*, 1–9. http://doi.org/10.1109/EMAP.2012.6507859.

Coit, D. W., Jackson, B. T. & Smith, a. E. (1998). Static neural network process models: Considerations and case studies. *International Journal of Production Research, 36*(11), 2953–2967. http://doi.org/10.1080/002075498192229.

Coit, D. W. & Smith, A. E. (1995). Using Designed Experiments to produce Robust Neural Network Models of Manufacturing Processes. In *Fourth Industrial Engineering Research Conference (IERC)*, May 1995, (pp. 1–10).

Ding, H., Ume, I. C., Powell, R. E. & Hanna, C. R. (2005). Parametric Study of Warpage in Printed Wiring Board Assemblies. *IEEE Transactions on Components and Packaging Technologies, 28*(3), 517–524.

Driel, W. D. Van, Zhang, G. Q., JAnssen, J. H. J., Ernst, L. J., Su, F., Chian, K. S. & Yi, S. (2003). Prediction and verification of process induced warpage of electronic packages. *Microelectronics Reliability*, *43*, 765–774. http://doi.org/10.1016/S0026-2714(03)00057-X.

Epec. (2014). *Wave Soldering Defects, Poor Lead Solderability and Wetting on a Printed Circuit Board*.

Franken, K., Maier, H. R., Prume, K. & Waser, R. (2000). Finite-Element Analysis of Ceramic Multilayer Capacitors : Failure Probability Caused by Wave Soldering and Bending Loads. *Journal of the American Ceramic Society*, *83*(6), 1433–1440.

Fu, C. Y. & Ume, C. (1995). Characterizing the temperature dependence of electronic packaging-material properties. *Jom*, *47*(6), 31–35. http://doi.org/10.1007/BF03221201.

Guo, Y. (1995). Application of shadow Moiré method in electronic packaging. In *1995 SEM Spring Conf. Exhibit, Grand Rapids, MI*, (pp. 12–14).

Hashimoto, T., Shin-ichiro, T., Morinishi, K. & Satofuka, N. (2008). Numerical simulation of conventional capillary flow and no-flow underfill in flip-chip packaging. *Computers & Fluids*, *37*(5), 520–523. http://doi.org/10.1016/j.compfluid.2007.07.007.

Hon, R., Lee, S. W. R., Zhang, S. X. & Wong, C. K. (2005). Multistack flip chip 3D packaging with copper plated through-silicon vertical interconnection. *2005 7th Electronic Packaging Technology Conference*, *2*, 384–389. http://doi.org/10.1109/EPTC.2005.1614434.

Huang, C. Y. & Huang, H. H. (2012). Process optimization of SnCuNi soldering material using artificial parametric design. *Journal of Intelligent Manufacturing*, *25*(4), 813–823. http://doi.org/10.1007/s10845-012-0720-z.

Hutapea, P. & Grenestedt, J. L. (2004). Reducing Warpage of Printed Circuit Boards by Using Wavy Traces. *Journal of Electronic Packaging*, *126*(3), 282–287. http://doi.org/10.1115/1.1756591.

Hutapea, P., Grenestedt, J. L., Modi, M., Mello, M. & Frutschy, K. (2006). Prediction of microelectronic substrate warpage using homogenized finite element models. *Microelectronic Engineering*, *83*(3), 557–569. http://doi.org/10.1016/j.mee.2005.12.009.

IPC-A-610E. (2010). Supported Holes-solder-Vertical fill (A), Through-Hole Technology. In *IPC-A-610E Acceptability of Electronic Assemblies*, (pp. 43–44). Bannockburn, Illinois: IPC.

Johnson, A. (2014). *Lead-free soldering turns up the heat on PCB carriers, Norplex Micarta, High performance thermoset composites*.

Jong, W. R., Chen, Y. R. & Kuo, T. H. (2005). Wire density in CAE analysis of high pin-count IC packages: Simulation and verification. *International Communications in Heat and Mass Transfer*, *32*(10), 1350–1359. http://doi.org/10.1016/j.icheatmasstransfer.2005.05.012.

Khor, C. Y., Abdullah, M. Z., Ariff, Z. M. & Leong, W. C. (2012). Effect of stacking chips and inlet positions on void formation in the encapsulation of 3D stacked flip-chip package. *International Communications in Heat and Mass Transfer*, *39*(5), 670–680. http://doi.org/10.1016/j.icheatmasstransfer.2012.03.023.

Khor, C. Y., Abdullah, M. Z. & Che Ani, F. (2011). Study on the fluid/structure interaction at different inlet pressures in molded packaging. *Microelectronic Engineering*, *88*(10), 3182–3194. http://doi.org/10.1016/j.mee.2011.06.026.

Khor, C. Y., Abdullah, M. Z. & Leong, W. C. (2012). Fluid/structure interaction analysis of the effects of solder bump shapes and input/output counts on moulded packaging. *IEEE Transactions on Components, Packaging and Manufacturing Technology*, *2*(4), 604–616. http://doi.org/10.1109/TCPMT.2011.2174237.

Koch, V. E. (1998). Surface mount assembly of BGA and μBGA. *Soldering & Surface Mount Technology*, *10*(1), 32–36.

Kulkarni, V. M., Seetharamu, K. N., Azid, I. A., Aswataha Narayana, P. A. & Abdul Quadir, G. A. (2006). Numerical simulation of underfill encapsulation process based on characteritsic split method. *International Journal for Numerical Methods in Engineering*, *66*(10), 1658–1671.

Kuntz, M. & Menter, F. R. (2004). Simulation of Fluid-Structure Interactions in Aeronautical Applications. In *European Congress on*

Computational Methods in Applied Sciences and Engineering, ECCOMAS 2004, Jyvaskyla, (pp. 1–13).

Kuo, C. H., Hua, H. H., Chan, H. Y., Yang, T. H., Lin, K. S. & Ho, C. E. (2013). Interfacial reaction and mechanical reliability of PTH solder joints with different solder/surface finish combinations. *Microelectronics Reliability,* 53(12), 2012–2017. http://doi.org/10.1016/j.microrel.2013.03.002.

Lau, C.-S., Abdullah, M. Z., Ani, F. C. & Che Ani, F. (2012). Three-dimensional thermal investigations at board level in a reflow oven using thermal-coupling method. *Soldering & Surface Mount Technology,* 24(3), 167–182. http://doi.org/10.1108/ 0954091121 1240038.

Lau, C., Abdullah, M. Z. & Che Ani, F. (2012). Computational fluid dynamic and thermal analysis for BGA assembly during forced convection reflow soldering process. *Soldering & Surface Mount Technology,* 24(2), 77–91. http://doi.org/10.1108/09540911211214659.

Lee, M. (2000). Finite element modelling of printed circuit boards (PCBs) for structural analysis. *Circuit World,* 26(3), 24–29. http://doi.org/10.1108/03056120010322861.

Lee, N. C. (2002). Introduction to Surface Mount Technology. In *Reflow Soldering Processes and Troubleshooting : SMT, BGA, CSP and Flip Chip Technologies,* (pp. 1–18).

Leong, W. C., Abdullah, M. Z. & Khor, C. Y. (2012). Application of flexible printed circuit board (FPCB) in personal computer motherboards: Focusing on mechanical performance. *Microelectronics Reliability,* 52(4), 744–756. http://doi.org/10.1016/j.microrel. 2011.11. 003.

Liukkonen, M., Havia, E., Leinonen, H. & Hiltunen, Y. (2009). Application of self-organizing maps in analysis of wave soldering process. *Expert Systems with Applications,* 36, 4604–4609. http://doi.org/10.1016/j.eswa.2008.05.016.

Liukkonen, M., Havia, E., Leinonen, H. & Hiltunen, Y. (2011a). Expert system for analysis of quality in production of electronics. *Expert Systems with Applications,* 38(7), 8724–8729. http://doi.org/10.1016/j.eswa.2011.01.081.

Liukkonen, M., Havia, E., Leinonen, H. & Hiltunen, Y. (2011b). Quality-oriented optimization of wave soldering process by using self-organizing maps. *Applied Soft Computing*, *11*(1), 214–220. http://doi.org/10.1016/j.asoc.2009.11.011.

LoVasco, F. & Oien, M. A. (1983). *Process for controlling solder joint geometry when surface mounting a leadless integrated circuit package on a substrate*, US4878611 A.

Martin, T. (2014). *Wave soldering problems*.

Morris, J. & Szymanowski, R. (2008). *Effect of Contact Time on Lead-Free Wave Soldering*.

Pei, C. C. & Hwang, S. J. (2005). Three-Dimensional Paddle Shift Modeling for IC Packaging. *Journal of Electronic Packaging*, *127*(3), 324. http://doi.org/10.1115/1.1938986.

Pietraszek, J., Gadek-Moszczak, A. & Torunski, T. (2014). Modeling of errors counting system for PCB soldered in the wave soldering technology. *Advanced Materials Research*, *874*, 139–143.

Polsky, Y., Sutherlin, W. & Ume, I. C. (2000). A Comparison of PWB Warpage Due to Simulated Infrared and Wave Soldering Processes. *IEEE Transactions on Electronic Packaging Manufacturing*, *23*(3), 191–199.

Ramdan, D., Abdullah, Z. M., Mujeebu, M. A., Loh, W. K., Ooi, C. K. & Ooi, R. C. (2011). *FSI Simulation of Wire Sweep PBGA Encapsulation Process Considering Rheology Effect*, 1–11.

Shen, Y. K., Huang, S. T., Chen, C. J. & Yu, S. (2006). Study on flow visualization of flip chip encapsulation process for numerical simulation. *International Communications in Heat and Mass Transfer*, *33*(2), 151–157.

Shi, X. Q., Zhou, W., Pang, H. L. J. & Wang, Z. P. (2014). Effect of Temperature and Strain Rate on Mechanical Properties of 63Sn/37Pb Solder Alloy. *Journal of Electronic Packaging*, *121*(September 1999), 179–185.

Suganuma, K., Ueshima, M., Ohnaka, I., Yasuda, H., Zhu, J. & Matsuda, T. (2000). Lift-off phenomenon in wave soldering. *Acta Materialia*, *48*(18–19), 4475–4481. http://doi.org/10.1016/S1359-6454(00)00234-2

Tan, W., & Ume, I. C. (2012). Application of Lamination Theory to Study Warpage Across PWB and PWBA During Convective Reflow Process. *IEEE Transactions on Components, Packaging and Manufacturing Technology*, *2*(2), 217–223. http://doi.org/10.1109/ TCPMT.2011. 2174793.

Teng, S. Y. & Hwang, S. J. (2008). Simulations and experiments of three-dimensional paddle shift for IC packaging. *Microelectronic Engineering*, *85*(1), 115–125. http://doi.org/10.1016/j.mee.2007.04. 148.

Tsai, T. N. (2009). Modeling and Optimization of Reflow Thermal Profiling Operation: A Comparative Study. *Journal of the Chinese Institue of Industrial Engineers*, *26*(6), 480–492. http://doi.org/10.1080/ 10170660909509162.

Ume, I. C., Martin, T. & Gatro, J. T. (1997). Finite element analysis of PWB warpage due to the solder masking process. *IEEE Transactions on Components, Packaging, and Manufacturing Technology: Part A*, *20*(3), 295–306. http://doi.org/10.1109/95.623024.

Wan, J. W., Zhang, W. J. & Bergstrom, D. J. (2009). Numerical modeling for the underfill flow in flip-chip packaging. *IEEE Transactions on Components and Packaging Technologies*, *32*(2), 227–234. http:// doi.org/10.1109/TCAPT.2009.2014355.

Wang, H., Zhou, H., Zhang, Y. & Li, D. (2010). Stabilized filling simulation of microchip encapsulation process. *Microelectronic Engineering*, *87*(12), 2602–2609. http://doi.org/10.1016/j.mee.2010. 07.026.

Wang, J., Wei, X., Zhu, W., Wu, J. & Wu, N. (2013). Study on Low Silver Sn-Ag-Cu-P Alloy for Wave Soldering. In *2013 20th IEEE International Symposium on the Physical and Failure Analysis of Integrated Circuits (IPFA)*, (pp. 485–489).

Wong, E. H. & Wong, C. K. (2009). Approximate solutions for the stresses in the solder joints of a printed circuit board subjected to mechanical bending. *International Journal of Mechanical Sciences*, *51*(2), 152–158. http://doi.org/10.1016/j.ijmecsci.2008.12.003.

Ying, M., Chia, Y. C., Mohtar, A., Yin, T. F. & Chuah, S. P. (2006). Thermal Induced Warpage Characterization for Printed Circuit Boards with Shadow Moiré System. In *8th Electronics Packaging Technology Conference EPTC '06., 6-8 Dec. 2006, Singapore*, (pp. 265–270).

In: Advances in Aerospace Science ...   ISBN: 978-1-53615-689-8
Editors: Parvathy Rajendran et al.   © 2019 Nova Science Publishers, Inc.

Chapter 2

KNITTED STRUCTURES IN AEROSPACE APPLICATIONS

Habib Awais and Mohd Shukur Zainol Abidin[*]
School of Aerospace Engineering, Universiti Sains Malaysia,
Pulau Pinang, Malaysia

ABSTRACT

Textile fabrics that were traditionally used for clothing purposes are currently being adopted in technical applications. Various techniques, including weaving and knitting, are used to develop textile fabrics. Knitting is the second most frequently used technique to manufacture fabrics. Warp and weft knitting are two types of knitting differentiated by the base of the yarn-feeding direction. The basic concept of knitting is discussed in this paper. The applications of knitted structures, namely, multiaxial warp-knitted and 3D spacer fabrics, in the aerospace industry are elaborated.

Keywords: weft knitting, warp knitting, spacer fabric, multiaxial fabric, knitted composite

---

[*] Corresponding Author's Email: aeshukur@usm.my.

1. Introduction

Textile fabrics are often utilised to fulfil the basic need of humans for clothing and for functional and aesthetic purposes. Various methods are used to manufacture textile fabrics from different raw materials, among which weaving and knitting are the prominent methodologies. Fabric formation methods or techniques exert significant impacts on the mechanical, physical, aesthetic and functional properties of fabrics.

Knitting is the second most commonly used method to produce textile fabrics, and it is rapidly becoming an active research area. The word knitting originated from the Dutch word 'knutten', which means 'to knot', and the word 'cnyttan' used as early as 1492 in English literature which also means 'to knot'. Interloping of yarns occurs during the knitting process. Such interloping can be carried out either by hand or machine, but the principles remain the same (i.e., pulling a new loop through the old loop). A pair of socks with cross stitch construction was found in Roman Egypt dating back to approximately the fifth century AD. Hand knitting was used in a 1350 painting of Northern Italy. Subsequently, Britain established cap knitting in 1424 (Nawab, Hamdani, & Shaker, 2017). Reverend William Lee was the inventor of hand frame knitting in 1589 and laid the foundation of today's modern knitting.

Knitting is better than weaving in terms of process capability, cost, flexibility and performance but is less durable. Knitting is further categorised into two broad categories (i.e., weft and warp knitting) depending on the yarn feeding and its direction (Spencer, 2001).

2. Weft Knitting

Weft knitting is more common and diverse than warp knitting due to its low capital cost, small space requirement, versatility, high production, low stock requirements for yarns and rapid pattern-changing facilities (Ashraf et al., 2015). If the direction of yarn feeding is perpendicular to the direction of fabric formation, then this type of knitting is called weft knitting (Figure

1), and the resultant fabric is known as weft-knitted fabric. The loops in weft knitting are generally formed across the fabric width. Weft-knitted fabrics are produced by the up and down movement of needles controlled by a cam system. Weft knitting is mainly used to make apparel, upholstery, furnishing and certain industrial fabrics (Spencer, 2001).

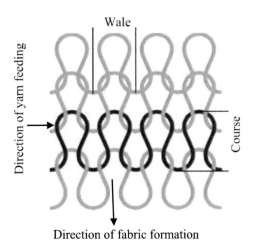

Figure 1. Weft knitting (Kumar & Vigneswaran, 2013).

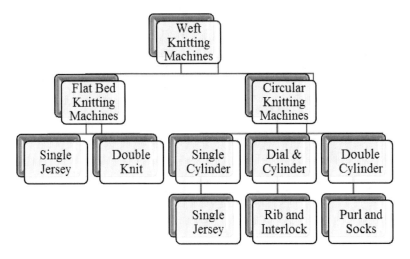

Figure 2. Classification of weft knitting machines (Nawab, 2016).

Weft knitting machines can be classified depending on the needle bed, machine diameter and shape, nature of the driving system, knitted structures, design elements and special products. A simple classification of weft knitting machines and their structures is shown in Figure 2.

Tubular fabric is generally produced on circular knitting machines. Circular knitting machines are divided into three major categories based on the cylinder and dial. Machines that only have a single cylinder and in which loop formation is carried out by the up and down movement of the needles inside this cylinder are called single-jersey knitting machines. Machines that have a dial and a cylinder and in which loop formation is carried out by the up and down movement of the cylinder needles coupled with the to-and-from movement of dial needles are known as rib and interlock machines. The difference between rib and interlock machines is the position of the dial and cylinder needles. The needles of the dial and cylinder are staggered in position in rib machines but perpendicular in the case of interlock machines, and this positioning is called rib gating and interlock gating, respectively (Figure 3).

Machines with double cylinders that are superimposed are called purl machines. The fabrics made by these knitting machines vary in terms of appearance and properties (Table 1), although the same structures are maintained. Therefore, a proper machine must be selected for a specific end use. Although the materials, structure and machine gauge can be changed, the selection of an accurate machine remains vital in achieving the required properties.

2.1. Knitting Machine Elements

The three main knitting elements are as follows:

- Needle
- Sinker
- Cam

Knitted Structures in Aerospace Applications

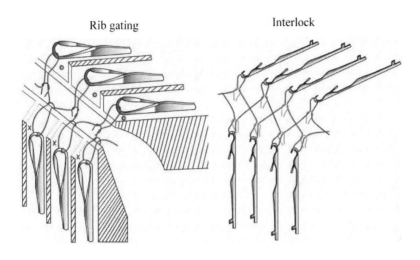

Figure 3. Needle gating (Spencer, 2001).

Table 1. Comparison of various knitted structures (Horrocks & Anand, 2000)

Characteristic	Single Jersey	1×1 Rib	Interlock	1×1 Purl
Appearance	Different face and back	Same on both sides	Same on both sides	Same on both sides
Thickness	Thicker	Much thicker	Very much thicker	Very much thicker
Unravelling	Either end	Only from the last knitted end	Only from the last knitted end	Either end
Curling behaviour	Tends to curl	Does not curl	Does not curl	Does not curl
Extensibility • Length wise • Width wise	Moderate (10% – 20%) High (30% – 50%)	Moderate Very high (50% – 100%)	Moderate Moderate	Very high High
End uses	Ladies' stockings Men's and women's shirts Base fabric for coating	Socks Collar and cuffs Waist bands Men's outer and underwear	Underwear Trousers Shirts Sportswear	Children's clothing Heavy outerwear

The needle is considered the most essential part of loop formation. Needles are placed on tricks available on the cylinder and dial to move freely for loop formation. Three basic types of needles are utilised in knitting machines.

(i) Latch needle
(ii) Spring bearded needle
(iii) Compound needle

The latch needle is the most commonly used in knitting machines due to its self-acting nature. The opening and closing processes are performed by the yarn itself. No auxiliary part is required for the opening and closing compared with the two other types of needles (Figure 4).

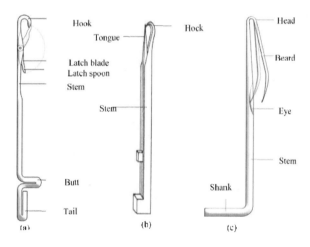

Figure 4. Parts of needle (a) Latch needle (b) Compound needle (c) Spring bearded needle (Spencer, 2001).

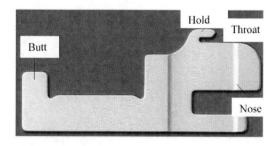

Figure 5. Parts of a sinker (Nawab, 2016).

The second most important element of knitting machines is the sinker, which is a thin metal plate placed perpendicular to the needle (Figure 5). The primary function of the sinker is to hold the old loop during and after clearing from the needle. The sinker also helps in loop formation and knocking over.

The sinker is used in weft and warp knitting machines, although the types of sinkers vary according to respective machines and depending on the required products. The function of the throat is to hold the old loop during loop formation, whilst the nose or belly is the projected portion in which the old loops and fabric stay. The butt is the part of the sinker that controls the to and from movements of the sinker along the horizontal direction via the cams. The hold or neb prevents yarn movement alongside the needle movement.

The third knitting element is the cam (Figure 6), which is a metal plate with channels on its surface. These channels make a path known as the cam track for needle movement during loop formation. The needle butt moves through the cams, and the profile of these cams decides the shape and length of the loop. Various types of cams are used in knitting. Three basic types of cams (i.e., knit, tuck and miss) are used in weft knitting, and they are known as stitch cams.

2.2. Loop Formation Cycle

An important and interesting parameter is the understanding of the loop formation cycle during knitting. All three types of needles perform a similar loop formation cycle to produce knitted fabric. The sole difference is the working of the parts of the needle that closes the needle for clearing. For a simplified understanding, the loop formation cycle with a latch needle is explained. Five stages are involved in the loop formation cycle (Figure 7).

(i) At the start, the needle is at rest position, and the latch is closed with the old loop inside the latch. This position is known as the idle position. The needle then starts to move upwards, the latch opens and the old loop starts to slip onto the stem. This is the run-in stage.

(ii) Afterwards, the needle continues to move upwards and attain its maximum height. The latch is now fully opened, and the old loop reaches the stem. This is known as the clearing stage.

(iii) In the yarn feeding stage, the needle starts descending, and a new yarn is engaged by the needle. At the same time, the old loop starts moving upwards to close the latch.

(iv) The fourth stage is the knock over or cast off. In this stage, the old loop closes the latch and moves in the outward direction to slide off.

(v) Loop pulling is the final stage of loop formation. During loop pulling, the needle reaches the lower-most position, and the new loop is pulled out from the old loop. The newly formed loop acts as the old loop for the next cycle. In this way, the loops are interconnected with one another, and the knitted fabric is produced.

3. WARP KNITTING

Warp knitting is commonly used to develop fabrics for structural purposes. In warp knitting, the direction of yarn feeding is parallel to the direction of fabric formation (Figure 8). The process of yarn feeding in warp knitting is identical to the weaving process through the warp beams. Each yarn passes through a guide and is fed into the individual needles to produce the fabric. The movement of the guide bar creates a variety of warp-knitted structures. The machines used for warp knitting are flat, complex and more diverse than weft knitting machines.

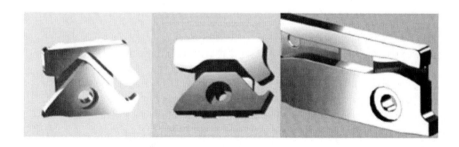

Figure 6. Stitch cams (Nawab et al., 2017).

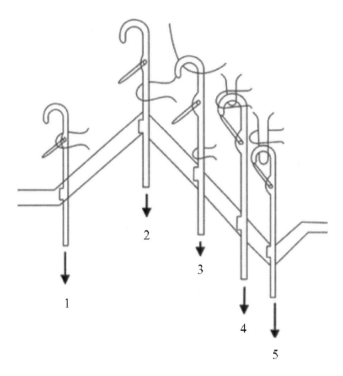

Figure 7. Loop formation cycle (Nawab, 2016).

Warp knitting machines are further classified into two categories, namely, tricot and raschel. These machines have many differences. A latch needle is used in raschel, whereas spring bearded, and compound needles are used in tricot machines. The sinkers are activated throughout the knitting cycle in tricot; in raschel, the sinkers work only during the rise of the needle. The tricot machine speed is 3500 courses per minute, whereas raschel machines have a lower speed of approximately 2000 courses per minute. Given that a smaller number of warp beams are used in tricot, comparatively simpler structures are produced by it. The angle between the needle and fabric take down in tricot machines is 90°, whereas it is 160° in raschel machines. The reduced number of warp beams and guide bars used in tricot results in cheaper machines than raschel machines (Spencer, 2001). Warp and weft knitting are commonly used in the textile industry to produce desired knitted fabrics, although both methods have various pros and cons.

4. APPLICATIONS OF KNITTED FABRICS

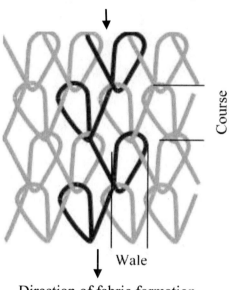

Figure 8. Warp knitting (Kumar & Vigneswaran, 2013).

Table 2. Comparison of weft and warp knitting (Nawab, 2016)

Parameters	Weft Knitting	Warp Knitting
Loop formation	Course wise/horizontal	Wale wise/vertical
Needles working	Sequential	Combined
Yarn supply	Cones	Warp beams
Preparatory process	Less	More
Production	Low	High
Stretch capability	High	Low
Dimensional stability	Low	High
Machine cost	Low	High
Yarns	Mostly staple	Mostly filament
Applications	T-shirts, jackets, suits, skirts, seamless hosiery, collar and cuffs, sweaters, etc.	Sportswear lining, track suits, furnishings, laundry bags, nets, shoe lining, car cushions, sun shades, production masks, etc.

Apart from the conventional use in apparel, knitted fabrics are widely used in structural applications, such as automotive, geotextile, medical textile and aerospace. As for aerospace applications, warp-knitted structures, namely, multiaxial warp-knitted fabrics and 3D weft-knitted structures, are suitable for structural requirements.

4.1. Multiaxial Warp-Knitted Fabrics

Multiaxial warp-knitted fabrics have gained a significant status in the field of industrial composites over the past few years. They were developed in the early '80s but have since made their way into structural composites by the '90s. Such fabrics are commonly produced on raschel machines but can also be produced on specialised tricot machines. These fabrics generally consist of two weft layers: one warp yarn and one weft yarn (Figure 9). The layers are stitched from both sides by pillar stitches. The diagonal weft insertion helps in the placement of the yarns at a constant distance.

Multiaxial knitted fabrics contain one or more fibre layers on top of one another and are held together by a binding yarn (typically polyester). To achieve the desired properties, several layers of fibres (glass, carbon, etc.) are stacked in different orientations. Multiaxial warp-knitted structures are characterised based on the direction of the reinforcement yarns. Commercially available fabrics are biaxial, triaxial and quadriaxial.

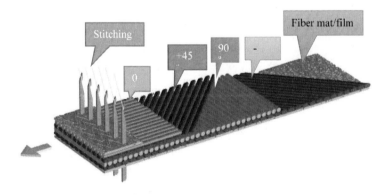

Figure 9. Schematic diagram of the multiaxial knitted fabric development (Hu, 2008).

Multiaxial knitted fabrics were originally produced as substitutes to flexible coated fabrics, but they soon gained importance in industrial applications due to their excellent mechanical properties. The stability and safety of such structures are crucial to the increase in applications. The mechanical behaviour of these structures, especially tensile properties, must be understood. Wang (2002) studied the mechanical behaviour of multiaxial warp-knitted composites using biaxial, triaxial and quadriaxial fabrics as reinforcements (Table 3).

Table 3. Mechanical characterisation of multiaxial warp-knitted composites (Wang, 2002)

Fabric	Volume fraction (%)	Direction	Compression E (GPa)	σ (MPa)	Flexural E (GPa)	σ (MPa)
Biaxial	47.8	0°	7.5	110	10.9	251
		45°	15.7	377	19.0	380
		90°	7.7	114	11.2	257
Triaxial	49.7	0°	11.6	270	16.4	423
		45°	14.3	353	17.2	503
		90°	12.2	256	14.9	440
Quadriaxial	41.5	0°	12.1	313	15.0	356
		45°	12.0	323	12.9	384
		90°	11.4	306	14.8	368

Multiaxial warp-knitted fabrics play a significant role in composites due to their low cost, good mechanical properties and design flexibility. Composites were initially used for secondary non-load-bearing assemblies, but the current trend is to use them in primary load-bearing structures that require improved mechanical properties (Du & Ko, 1996). Composites made of multiaxial warp-knitted fabrics are now being used in military, aerospace, and commercial aircraft industries because of their low weight, high strength, flexibility and customizable physical and mechanical properties. A number of aircraft use these composites for critical structures as the fuselage and wings. Aircraft skin is the most recent application of such composites. The fuselage and engine panelling, top and side tail units, leading edges on the rudders, rotor blades and ballistic protection in helicopters are the main applications of multiaxial warp-knitted fabric composites. Approximately 20% of multiaxial warp-knitted composites are used in the aerospace industry (Hu, 2008).

4.2. Weft-Knitted Fabrics

4.2.1. Aircraft Panel Structures

Although weft-knitted fabrics have low in-plane mechanical properties, these properties can be increased by using inlay yarns. Weft-knitted fabrics are utilised as structural parts of an aeroplane (i.e., wings), particularly as preforms. Several researchers have used 3D weft-knitted structures for the development of aeroplane wings. For this purpose, the 3D knitted preform was developed using a Stoll flat knitting machine. Afterwards, the preforms were infused with the matrix resin to produce composites via the RTM technique (Costa et al., 2002).

Complex 3D spacer fabrics are developed by using two separate layers of fabric vertically connected by a pile of yarns or fabric layers by using weaving and knitting technologies. 3D spacer fabrics with multilayer reinforcement exhibit superior mechanical properties and are suitable for lightweight composite applications as a replacement of the conventional metal panel structures of aircraft (Abounaim & Cherif, 2012). Aircraft push rod fairing is manufactured by a flat knitting machine using glass fibre (Kumar & Vigneswaran, 2013). Spacer fabrics can be developed via weft and warp knitting.

4.2.2. Electromagnetic Shielding

Electromagnetic waves from electronic and telecommunication devices can potentially exert effects on the human body. Therefore, electromagnetic shielding materials are used in industrial applications to reduce or eliminate unwanted radiation or signals (Cheng, Ramakrishna, & Lee, 2000). Polymeric composites, such as knitted-fabric-reinforced polypropylene composites, are extensively used for this purpose. Copper is used in these composites as conductive filaments. Interlock fabrics have better shielding effectiveness from low to high frequency than plain and rib structures. Copper-core-knitted fabrics are currently used to shield the electronic systems of aeroplanes (Gokarneshan, Varadarajan, Sentil kumar, Balamurugan, & Rachel, 2012).

CONCLUSION

The demand for textile preforms for structural applications is increasing due to the good formability and low cost of these preforms. An important preform is knitted fabric. Knitting is the most flexible fabric formation technique used to control individual yarns through different types of stitches. This flexibility helps in controlling directional mechanical properties under specific load conditions. Various structures can be produced by knitting technology in 2D and 3D forms, which make it suitable for aerospace, automotive and other structural applications. An overview of knitting technology and knitted structures (especially multiaxial warp-knitted ones) and the benefits and limitations of knitting technology are highlighted in this chapter.

ACKNOWLEDGMENT

The authors gratefully acknowledge National Textile University, Pakistan for their funding under Faculty Development Program. This study was also funded by Universiti Sains Malaysia Short Term Grant 304/PAERO/6315057.

REFERENCES

Abounaim, md. & Cherif, C. (2012). Flat-knitted innovative three-dimensional spacer fabrics: A competitive solution for lightweight composite applications. *Textile Research Journal*, *82*(3), 288–298. http://doi.org/10.1177/0040517511426609.

Ashraf, W., Nawab, Y., Maqsood, M., Khan, H., Awais, H., Ahmad, S. & Ahmad, S. (2015). Development of seersucker knitted fabric for better comfort properties and aesthetic appearance. *Fibers and Polymers*, *16*(3), 699–701. http://doi.org/10.1007/s12221-015-0699-0.

Cheng, K. B., Ramakrishna, S. & Lee, K. C. (2000). Electromagnetic shielding effectiveness of copper/glass fiber knitted fabric reinforced polypropylene composites. *Composites Part A: Applied Science and Manufacturing*, *31*(10), 1039–1045. http://doi.org/10.1016/S1359-835X(00)00071-3.

Costa, A. N. N., Novo, C., Correia, N., Marques, A. T., De Araújo, M., Fangueiro, R. & Ciobanu, L. (2002). Structural Composite Parts Production from Textile Preforms. *Key Engineering Materials*, *230–232*, 36–39. http://doi.org/10.4028/www.scientific.net/KEM.230-232.36.

Du, G. W. & Ko, F. (1996). Analysis of multiaxial warp-knit preforms for composite reinforcement. *Composites Science and Technology*, *56*(3), 253–260. http://doi.org/10.1016/0266-3538(95)00108-5.

Gokarneshan, N., Varadarajan, B., Sentil kumar, C. B., Balamurugan, K. & Rachel, A. (2012). Engineering knits for versatile technical applications: Some insights on recent researches. *Journal of Industrial Textiles*, *42*(1), 52–75. http://doi.org/10.1177/1528083711426021.

Horrocks, A. R. & Anand, S. C. (2000). *Handbook of Technical Textiles*. Woodhead Publishing Limited. http://doi.org/10.1533/9781855738966.372.

Hu, J. (2008). *3-D fibrous assemblies: Properties, applications and modelling of three-dimensional textile structures*. 感染症誌, (Vol. 91). Woodhead Publishing Limited.

Kumar, R. S. & Vigneswaran, C. (2013). *Textiles for Industrial Applications*. http://doi.org/10.1191/0269215505cr893oa.

Nawab, Y. (2016). *Textile Engineering: An Introduction*. De Gruyter Oldenbourg.

Nawab, Y., Hamdani, S. T. A. & Shaker, K. (2017). *Structural Textile Design*. CRC Press. http://doi.org/10.1201/9781315390406.

Spencer, D. J. (2001). *Knitting Technology* (3rd ed.). Woodhead Publishing Limited.

Wang, Y. (2002). Mechanical Properties of Stitched Multiaxial Fabric Reinforced Composites From Mannual Layup Process. *Applied Composite Materials*, *9*(2), 81–97.

In: Advances in Aerospace Science ...　　ISBN: 978-1-53615-689-8
Editors: Parvathy Rajendran et al.　　© 2019 Nova Science Publishers, Inc.

Chapter 3

CARBON NANOTUBE-REINFORCED HIERARCHICAL CARBON FIBRE COMPOSITES

Mohd Shukur Zainol Abidin[*]
School of Aerospace Engineering, Universiti Sains Malaysia,
Pulau Pinang, Malaysia

ABSTRACT

The application of polymer composites as primary structures is limited by the sensitivity of these materials to damage and defects. Although polymer composites exhibit excellent in-plane tensile performance, they suffer from relatively poor compression, interlaminar and other matrix-dominated properties. Approaches to address these concerns include through-thickness reinforcement and advanced toughening of matrix resins. However, these approaches are expensive and sensitive to processing conditions; thus, they frequently lead to reductions in the pristine and fatigue performance of composite materials. Recently, the inclusion of carbon nanotubes (CNTs) in the matrix of fibre-reinforced composites to produce hierarchical composites has been identified as a promising means of improving damage tolerance without compromising

---

[*] Corresponding Author's Email: aeshukur@usm.my.

pristine in-plane properties. This chapter explores the mechanical enhancements observed particularly in matrix-dominated properties, such as interlaminar shear strength, flexural properties and interlaminar fracture toughness, to highlight the enhancement obtained from the incorporation of CNTs into the composite material system.

Keywords: hierarchical composites, carbon nanotubes, carbon fibre

1. INTRODUCTION

Several of the major drawbacks in employing carbon fibre-reinforced polymer (CFRP) composites in structural applications are the relatively poor compression performance and interlaminar toughness of these composites. These mechanical properties are dominated by matrix and interphase properties, which are often orders of magnitude inferior to fibre-dominated properties, thus making CFRP composites susceptible to premature and unexpected failures when loaded under compression or subjected to delamination damage. Employing composite materials in critical load-bearing structures must therefore be supported by extensive verification and testing.

Significant studies have been conducted to improve the damage tolerance of composite materials for structural applications. Through-thickness reinforcements, such as laminate stitching, tufting and z-pinning, have been extensively investigated and demonstrated to significantly improve fracture toughness, impact resistance, damage tolerance and stiffener pull-off strength. However, these through-thickness reinforcements often damage the primary fibre reinforcements, thus reducing the pristine in-plane properties of composites.

Another method to improve damage tolerance is toughening the matrix constituent. This approach can be implemented by doping the matrix with thermoplastic or elastomer particles. The modified matrix would then potentially possess increased fracture energy absorption capacity by failing via favourable plastic deformation. However, the in-plane properties of toughened composites are often compromised. An alternative modification to this toughening method is doping composites with nanofillers, such as nanoclays, nanofibres and CNTs, which possess high strength and stiffness. This technique has been demonstrated to improve damage tolerance without losses in pristine in-plane properties and will be discussed further in this review.

2. CARBON NANOTUBES (CNTS)

Since the work of Iijima, who observed and reported the properties of CNTs, many studies have investigated the physical, electrical and mechanical properties of this new form of carbon. As with all nanostructured materials, the properties of nanostructured composites are highly structure- and size-dependent. Single-walled CNTs (SWCNTs) are fullerene-based structures that can be considered a cylinder of carbon graphite (at 1 nanometre diameter). Multi-walled carbon nanotubes (MWCNTs) are like SWCNTs but have many layers in the cylindrical structure (5–50 nanometre in diameter) (Figure 1).

Several methods can be used to synthesise CNTs. Iijima utilised arc vaporisation to produce the first images of MWCNTs. A similar method is laser evaporation, which can produce SWCNTs. However, these methods require a large amount of energy with a low product yield. The third method, catalytic chemical vapour deposition (CVD), is a much-preferred method for industrial-scale CNT production.

CNTs are amongst the stiffest and strongest materials ever made. In theory, their Young's modulus reaches 1000–1200 GPa, and their experimental tensile strengths range from 11–63 GPa. Although the reported experimental values vary widely, the consensus is that CNTs possess higher mechanical properties than most metals. Furthermore, CNTs have attractive electrical properties that have been shown to be metallic or semiconducting depending on their structure and diameter. CNTs have also been experimentally demonstrated to possess very high thermal conductivities.

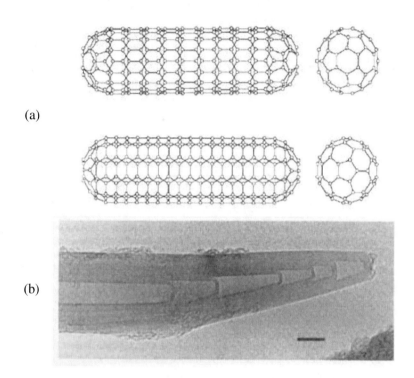

Figure 1. Different types of carbon nanotubes: (a) schematic of SWCNTs and (b) images of MWCNT cap (5 nm scale bar) (Iijima, 1991).

3. HIERARCHICAL COMPOSITES (HCs)

The matrix-dominated properties in conventional fibre-reinforced composites can potentially benefit from the inclusion of CNTs that possess remarkable mechanical properties. Such multi-level reinforcements, which are known as multi-scale, hybrid or HCs, occur in nature. These HCs possess high mechanical performance despite having been formed from weak constituents.

3.1. Processing Methods for HCs

The combination of nanoparticles and conventional polymer composites to produce HCs can be achieved via three routes.

i. Growing or grafting nanoparticles onto the primary fibre constituent
ii. Dispersion of nanoparticles into the bulk matrix
iii. Selective reinforcement on prepreg interlaminar surfaces

Nanocomposites (NCs) are produced after CNTs have been dispersed into the matrix. A critical issue in NCs is ensuring a uniform dispersion of nanoparticles within the matrix and resulting composites. CNTs form agglomerates because of the van der Waals interaction between particles and their high surface energy. Many methods, such as separation in solvents followed by ultrasonic agitation, have produced good dispersion results. However, excessive external work can damage the surfaces of CNTs. Thostenson et al. (2005) also recognised the problem of aligning CNTs in the matrix because CNTs have a very small diameter but a very large aspect ratio.

NCs can be prepared through several production methods. The simplest method is mixing nanotubes in a polymer solution and removing the solvents afterwards. Further modification can be performed by acid treatment, functionalisation of CNTs and addition of a surfactant to the CNT–polymer solution. A problem with this method is that CNTs agglomerate during processing and therefore prevent good dispersion in the matrix. Furthermore, the content of CNTs that can be dispersed in the solvent is limited to avoid agglomeration. Direct mixing using mechanical, shear or ultrasonic techniques is also commonly used to add CNTs to low-viscosity thermosetting resins. Variations of this technique involve using three roll mills. However, the maximum CNT content is still limited because CNT inclusions severely increase the processing temperature and viscosity of NCs. A novel method in which a solid thermosetting epoxy/CNT mixture is treated as a highly viscous material and processed via high shear melt mixing is under investigation. This method could process high contents of CNTs (up to 25.0 wt%) with good dispersion.

Once the dispersion of CNTs in the matrix is achieved, the resulting modified matrix is combined with the primary reinforcement constituents. Conventional resin transfer moulding (RTM) and vacuum-assisted RSM have been used to infiltrate fibres with liquid-modified resins. However, the maximum content of CNTs that can be included into the matrix using liquid-modified resins is limited because CNTs dramatically increase the viscosity of the resulting matrix. Hence, most reported work used less than 2 wt% loading fractions. Another method of producing HCs is the filament winding technique that impregnates the fibre with the resin.

3.2. Mechanical Properties of HCs

Many studies have shown that fibre-dominated properties, such as tensile and flexure properties, remain relatively unaffected. The primary interest of HCs is in the matrix-dominated properties, such as off-axis tensile, in-plane compression, shear, and interlaminar fracture toughness. The expected improvements in compressive and interlaminar toughness after the incorporation of nanofillers into the matrix are illustrated in Figure 2.

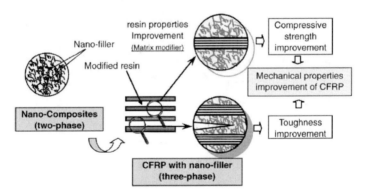

Figure 2. Schematic of mechanical property improvement of CFRPs by incorporation of nanofillers (Yokozeki, Iwahori, & Ishiwata, 2007).

3.2.1. Interlaminar Shear Properties

Significant interlaminar shear strength (ILSS) improvements are observed after the inclusion of nanoparticles in the composite. ILSS is a mechanical property often governed by fibre/matrix interphase performance. Consequently, improvements in the fibre/matrix interphase lead to an increase in ILSS. Significant improvements have been reported even with the inclusion of a small weight percentage of nanoparticles. However, several studies did not include the weight fraction of nanoparticles. Therefore, the direct link between nanofiller content and property improvement cannot be deduced. The reported improvement in the ILSS of HCs is listed in Table 1. The short beam shear (SBS) test is the most common testing method. In this test, interlaminar shear is generated indirectly via bending, and the short beam shear strength is often regarded as the ILSS of the composite. This test only requires small test pieces compared with other methods, and the test setup is simple.

A strengthened interface between fibres and nanofillers has been proposed as a possible enhancement mechanism for increasing strength. A direct increase in the strength and stiffness of the matrix by the inclusion of CNTs increases the efficiency of stress transfer and reduces the modulus mismatch between the matrix and fibre. However, the interlaminar adhesion properties must be treated with caution because most of the reported results are based on low loadings of nanofillers. With a high nanofiller content, the agglomeration of fillers could prevent good adhesion, thus reversing the trend in mechanical properties.

Very few studies have been conducted on measuring the in-plane shear modulus of HCs. An improved shear stiffness property of HCs can be expected upon achieving good dispersion and adhesion with the inclusion of CNTs to the composites. Qi et al. (An et al., 2013) showed that a 30% increase in shear modulus can be achieved by grafting CNTs onto a woven glass fibre composite because of the efficient stress transfer within the HCs. However, Godara et al. (Godara et al., 2009) reported an almost 15% reduction in shear modulus in the composites after the inclusion of CNTs because of the probable presence of CNT agglomerates in the samples.

Table 1. ILSS improvement exhibited by HCs

Matrix	Fibre	Nano-filler type	Maximum nano-filler content	Maximum strength improve-ment over parent	Test method	Reference
Epoxy	GF	DWCNT	0.1 wt% 0.30 wt%	15.7% 19.8%	SBS	(Gojny et al.)
Epoxy	GF	DWCNT	0.3 wt%	16%	SBS	(Qiu et al.)
Epoxy	GF	MWCNT	1 wt%	7.9%	SBS	(Qiu et al.)
Epoxy	Woven GF	MWCNT	0.5 wt% 1 wt%	3.2% 18.2%	SBS	(Fan et al.)
Epoxy	Woven GF	Nanofibres	0.1 wt% 1 wt%	23% 8%	SBS	(Green et al.)
Epoxy	Carbon fibres	MWCNT	1 wt% 5 wt%	31.2% 45.6	Lap shear	(Hsiao et al.)
Vinyl ester	Woven GF	SWCNT	0.1 wt% 0.2 wt%	18.5% 45%	SBS	(Zhu et al.)
Epoxy	CF	MWCNT	0.25 wt%	27%	SBS	(Bekyarova et al.)
Epoxy	Woven alumina	Grown CNTs	Up to 2.5 wt%	69%	SBS	(Enrique J. Garcia, Wardle, John Hart, et al.)
Epoxy	CF	Silica	10 wt% 20 wt%	-3.1% -12.2%	SBS	(Tang et al.)
Epoxy	CF Fabric	MWCNT	3 vol%	117%	SBS	(Singh et al.)
Epoxy	Woven GF	MWCNT	1 wt%	16%	±45° Tensile	(Aguilar Ventura et al.)
Epoxy	Woven GF	MWCNT	14.2 vol%	80%	±45° Tensile	(An et al.)
Epoxy	Woven CF	MWCNT	10.1 vol%	70%	±45° Tensile	(An et al.)
Epoxy	Woven GF	CN-Al$_2$O$_3$ hybrids	-	11%	SBS	(Li et al.)

* DWCNT – double-walled carbon nanotubes.
* GF – glass fibre.

3.2.2. Flexural Properties

The reported flexural properties of HC are listed in Table 2. Fibre-dominated properties, such as flexural stiffness, are not expected to be greatly influenced by CNT inclusions. Furthermore, modifying the fibre surface via grafting or CNT growth could damage the pristine fibre, resulting in an inferior composite compared with the parent composite.

Composites under flexural load could exhibit three types of failure modes, namely, tensile fibre failure, compression failure or mid-plane delamination. Compressive failure and mid-plane delamination are matrix-dominated properties. Hence, the inclusion of CNTs is expected to improve flexural strength if such failure modes are observed in HCs.

However, the improvement is subject to whether or not a homogeneous CNT dispersion is achieved in the HC. Entanglements of CNTs could create agglomerates, which are flaws that are detrimental to flexural strength. Flexural property improvements over the parent composites are attributed to matrix toughening, CNT bridging and mechanical interlocking.

Table 2. Flexural properties of HCs

Matrix	Fibre	Nano-filler type	Maximum nano-filler content	Maximum stiffness improvement	Maximum strength improvement	Test method	Reference
Epoxy	Woven CF	CSCNT	5 wt% 10wt%	6.3% 4.0%	13.9% 18.3%	3 pt. bending	(Iwahori et al.)
Epoxy	Woven CF	Forest grown CNT	2 wt%	5.2%	140%	3 pt. bending	(Veedu et al.)
Epoxy	Woven CF	MWCNT	2 wt%	2%	22.3%	3 pt. bending	(Yuanxin et al.)
Epoxy	Woven CF	MWCNT	0.3 wt%	4.9%	3.0%	3 pt. bending	(Yuanxin et al.)
Epoxy	CF	CSCNT	5 wt% 10 wt%	4.0% 5.3%	4.2% 1.5%	3 pt. bending	(Yokozeki, Iwahori, Ishiwata, et al.)
Phenolic	UD CF 2D CF Woven CF	Grown CNT	9.1 wt% 8.3 wt% 18.42 wt%	28% 54% 46%	20% 75% 66%	3 pt. bending	(Mathur et al.)
Epoxy	Woven GF	CNT–Al_2O_3 hybrids	-	19%	12%	3 pt. bending	(Li et al.)

Generally, improvements in flexural modulus and strength in UD systems are not significant, but improvements in flexural properties are significant in woven fibre systems. This result also shows that the surface-grown CNT method provides the highest flexural improvements whilst CNT dispersion in the matrix offers only slight improvements.

3.2.3. Delamination Fracture Toughness

Other important mechanical properties that have shown significant improvements with CNT inclusion are delamination resistance and crack tip arrest. The toughening mechanisms observed in nanocomposite fractures (Figure 3) are also manifested in HCs. The reported enhancements in the fracture toughness of HCs are listed in Table 3.

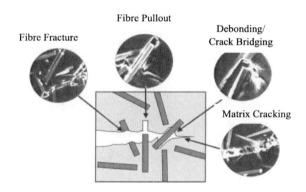

Figure 3. Key mechanisms of energy dissipation identified in the fracture of short and continuous fibre-reinforced composites (Thostenson et al., 2005).

HCs manufactured via the CNT growth/grafted method exhibit a significant improvement in fracture toughness whilst those manufactured with nanoparticles dispersed within the matrix present only modest improvements. However, the fracture toughness improvements in HCs using the modified interlaminar region method must be treated with caution because the modification could increase the resin-rich interply thickness. This increased thickness could in turn increase fracture toughness. Several other fracture mechanisms have been proposed to explain the increase in the fracture toughness of HCs. R. Sadeghian et al. (2006) and Karapappas et al. (2009) suggested that the improvements are due to the rough fracture surfaces that require more energy to propagate the crack tip. The CNT network structures in HCs that could lead to an improved stress transfer between the matrix and fibres are also attributed to the improvement in fracture toughness.

Table 3. Fracture toughness improvements of HCs over the parent composite

Matrix	Fibre	Nano filler type	Maximum nano-filler content	Maximum fracture toughness improvement	Test method	Reference
Epoxy	CF	Grown CNT	-	150% 300%	DCB ENF	(E. J. Garcia, Wardle, & Hart)
Epoxy	CF	DWCNT MWCNT	0.5 wt%	55% 83%	DCB	(Godara et al.)
Epoxy	CF	Grown CNT	-	358% 54%	DCB ENF	(Veedu et al.)
Epoxy	CF	MWCNT	0.1 wt% 0.3 wt% 0.5 wt% 1 wt%	4% 13% 9% 5%	DCB	(Romhany et al.)
Epoxy	Woven CF	Grown CNT	-	55%	DCB	(Kepple et al.)
Epoxy	GF	DWCNT	0.30 wt%	No improvement	DCB/ENF	(Wichmann et al.)
Polyester	GF	CNF	1 wt%	100%	DCB	(Ramin Sadeghian et al.)
Epoxy	CF	CSCNT	5 wt%	98% 30%	DCB ENF	(Yokozeki, Iwahori, Ishiwata, et al.)
Epoxy	CF	MWCNT	1 wt%	60% 75%	DCB ENF	(Karapappas et al.)
Epoxy	CF	SWCNT	0.1 wt%	13% 28%	DCB ENF	(Ashrafi et al.)
Epoxy	Woven CF	MWCNT	-	26% 38%	DCB ENF	(Joshi et al.)
Epoxy	Woven CF	MWCNT	10.1 vol%	~80%	DCB	(An et al.)
Epoxy	Alumina Fabric	Grown CNT	-	~100%	DCB	(Wicks et al.)
Epoxy	Woven CF	Grown CNT	-	~67% ~60%	DCB ENF	(Du et al.)
Epoxy	CF	MWCNT	0.3 wt%	143% 127%	DCB ENF	(Mirjalili et al.)

Furthermore, crack bridging by CNTs could blunt the crack tip during propagation. The inclusion of CNTs can be used to steer the crack path by adjusting the microfibre orientation, thus creating a tortuous path. However, whether the interactions between the fracture mechanisms co-exist or compete within a fractured sample remains ambiguous. Oftentimes, an increase in fracture toughness by the inclusion of CNTs also translates to high impact performance.

3.2.4. Enhancing the Fracture Toughness of Composites

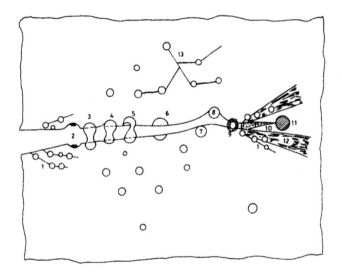

Figure 4. Crack toughening mechanisms in rubber-filled modified polymers: (1) shear band formation near rubber particles; (2) fracture of rubber particles after cavitation; (3) stretching, (4) debonding and (5) tearing of rubber particles; (6) transparticle fracture; (7) debonding of hard particles; (8) crack deflection by hard particles; (9) voided/cavitated rubber particles; (10) crazing; (11) plastic zone at the craze tip; (12) diffuse shear yielding; and (13) shear band/craze interaction (Garg et al., 1988).

Enhancing the fracture toughness of conventional fibre-reinforced polymer composites has been an active research area in the past decades. Kim and Mai (Kim et al., 1991) published an extensive review regarding the methodologies explored to improve the fracture toughness of polymer reinforced composites. One of the methodologies, namely, intrinsically toughening the matrix via the inclusion of thermoplastics or rubber-modified thermosets, has already been commercialised (e.g., HexPly® 8552, HexPly®M21 and HexPly®M91). The modified matrix promotes plastic yielding at the crack tip, which then absorbs and arrests the crack propagation, as illustrated in Figure 4.

Several of the mechanisms, such as crack deflection and crack pinning, also manifest in CNT-reinforced composites. However, toughening the matrix with thermoplastic or rubber particles often compromises the pristine in-plane mechanical and fatigue properties.

3.2.5. Toughness Improvement via Heterogeneity

Another interesting methodology to enhance fracture toughness is generating intermittent weak and strong interfacial bonding between the matrix and fibres. This method demonstrates an impressive 400% fracture toughness improvement compared with unmodified composites. Interfacial debonding occurs at the weak fibre/matrix interphase, and crack blunting by the Cook–Gordon mechanism occurs, as illustrated in Figure 5.

This intermittent bonding concept has been shown to improve the translaminar and interlaminar fracture toughness, suggesting that 'imperfection' in the composite can be beneficial if correctly engineered. The tortuosity of the crack tip can also be generated by modulating the length and orientation of the grafted CNTs on the fibres.

The fracture toughness improvements exhibited by this intermittent bonding methodology are attributed to increased tortuosity and great interaction of the crack tip with the matrix, fibre and fibre/matrix interphase. Furthermore, this methodology shows that pursuing perfection in processing might not be necessary and cost effective.

Such hierarchical morphologies are often exhibited by nature, such as in bones, abalone shells and crab exoskeletons. A homogenous and perfectly arranged structure rarely exists in nature. For example, the microstructure of abalone shells (Figure 6) consists of layers of aragonite tiles and is further reinforced by nanostructured organic materials between the layers. Mimicking the structures optimised by nature over significant evolutionary time might be beneficial in harnessing the untapped potential of a material.

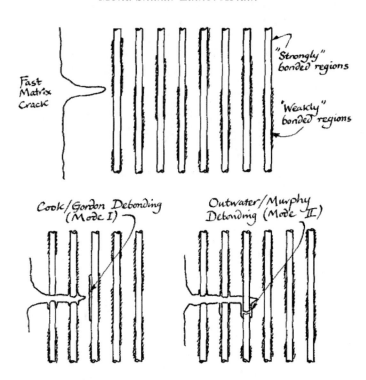

Figure 5. Fibres with alternated regions of engineered low interfacial strength and strong interfacial strength showing the debonding mechanisms (Atkins, 1975).

However, the critical intermittent length required for an optimised hierarchical system is unclear. A key aspect of this critical length is generating enough bonding region between the matrix and fibre, thereby ensuring that the Rule of Mixture strength is obtained. Furthermore, the process zone within the material is expected to dictate the stress state at the crack tip. A simple model to estimate the size of the process zone for a brittle material is suggested by Irwin's modifications (Irwin, 1957) (Equation 1).

$$r_y = \frac{1}{2\pi}\left(\frac{K}{\sigma_{ys}}\right)^2, \qquad (1)$$

where r_y is the radius of the process zone, K is the fracture toughness of the material and σ_{ys} is the yield tensile strength (Figure 7).

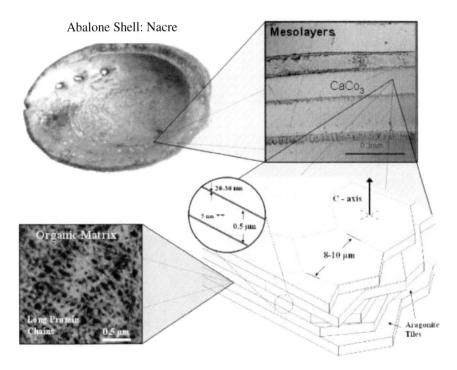

Figure 6. Hierarchical structure of an abalone shell (Meyers et al., 2008).

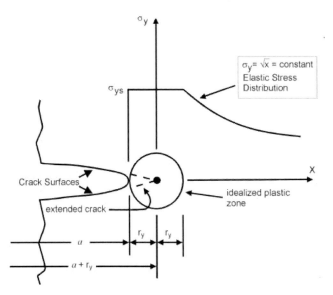

Figure 7. Schematic of the process zone at the crack tip (Miedlar et al., 2006).

Conclusion

Significant progress has been achieved in the study of HCs that demonstrate a significant enhancement in mechanical, electrical and thermal properties. Oftentimes, the reported improvements are exhibited with the inclusion of a relatively small quantity of CNTs. However, the properties of HCs with extremely high CNT loading (>20 wt%) remain inconclusive.

An approach to this methodology is to promote preferential mechanisms and tailor the morphology to achieve the desired performance. Such a methodology could provide an opportunity to increase the potential of nanoparticle inclusion in the system. By placing the mechanisms as the target, the CNT content, distribution, alignment, dispersion and processing can be engineered to achieve desirable properties.

Acknowledgments

The author acknowledges the support provided by Universiti Sains Malaysia Short Term Grant 304/PAERO/6315057.

References

Aguilar Ventura, & Lubineau. (2013). The effect of bulk-resin CNT-enrichment on damage and plasticity in shear-loaded laminated composites. *Composites Science and Technology, 84*(0), 23-30. doi: http://dx.doi.org/10.1016/j.compscitech.2013.05.002.

An, Rider, & Thostenson. (2012). Electrophoretic deposition of carbon nanotubes onto carbon-fiber fabric for production of carbon/epoxy composites with improved mechanical properties. *Carbon, 50*(11), 4130-4143. doi: http://dx.doi.org/10.1016/j.carbon.2012.04.061.

An, Rider, & Thostenson. (2013). Hierarchical composite structures prepared by electrophoretic deposition of carbon nanotubes onto glass fibers. *ACS Applied Materials & Interfaces, 5*(6), 2022-2032. doi: 10.1021/am3028734.

Ashrafi, Guan, Mirjalili, Zhang, Chun, Hubert, Simard, Kingston, Bourne, & Johnston. (2011). Enhancement of mechanical performance of epoxy/carbon fiber laminate composites using single-walled carbon nanotubes. [Article]. *Composites Science and Technology, 71*(13), 1569-1578. doi: 10.1016/j.compscitech.2011.06.015.

Atkins. (1975). Intermittent bonding for high toughness - high-strength composites. *Journal of Materials Science, 10*(5), 819-832. doi: 10.1007/bf01163077.

Bekyarova, Thostenson, Yu, Kim, Gao, Tang, Hahn, Chou, Itkis, & Haddon. (2007). Multiscale carbon nanotube - carbon fiber reinforcement for advanced epoxy composites. *Langmuir, 23*(7), 3970-3974. doi: doi:10.1021/la062743p.

Du, Liu, Xu, Zeng, & Mai. (2014). Flame synthesis of carbon nanotubes onto carbon fiber woven fabric and improvement of interlaminar toughness of composite laminates. *Composites Science and Technology, 101*(0), 159-166. doi: http://dx.doi.org/10.1016j.compscitech.2014.07.011.

Fan, Santare, & Advani. (2008). Interlaminar shear strength of glass fiber reinforced epoxy composites enhanced with multi-walled carbon nanotubes. *Composites Part A: Applied Science and Manufacturing, 39*(3), 540-554. doi: DOI 10.1016/j.compositesa.2007.11.013.

Garcia, Wardle, & Hart. (2008). Joining prepreg composite interfaces with aligned carbon nanotubes. *Composites Part A: Applied Science and Manufacturing, 39*(6), 1065-1070.

Garcia, Wardle, John Hart, & Yamamoto. (2008). Fabrication and multifunctional properties of a hybrid laminate with aligned carbon nanotubes grown *In Situ*. *Composites Science and Technology, 68*(9), 2034-2041. doi: 10.1016/j.compscitech.2008.02.028.

Garg, & Mai. (1988). Failure mechanisms in toughened epoxy resins - a review. *Composites Science and Technology, 31*(3), 179-223. doi: http://dx.doi.org/10.1016/0266-3538(88)90009-7.

Godara, Mezzo, Luizi, Warrier, Lomov, van Vuure, Gorbatikh, Moldenaers, & Verpoest. (2009). Influence of carbon nanotube reinforcement on the processing and the mechanical behaviour of carbon fiber/epoxy composites. *Carbon, 47*(Copyright 2010, The Institution of Engineering and Technology), 2914-2923.

Gojny, Wichmann, Fiedler, Bauhofer, & Schulte. (2005). Influence of nano-modification on the mechanical and electrical properties of conventional fibre-reinforced composites. *Composites Part A: Applied Science and Manufacturing, 36*(11), 1525-1535. doi: DOI 10.1016/j.compositesa. 2005.02.007.

Green, Dean, Vaidya, & Nyairo. (2009). Multiscale fiber reinforced composites based on a carbon nanofiber/epoxy nanophased polymer matrix: Synthesis, mechanical, and thermomechanical behavior. *Composites Part A: Applied Science and Manufacturing, 40*(9), 1470-1475. doi: 10.1016/j.compositesa.2009.05.010.

Hsiao, Alms, & Advani. (2003). Use of epoxy/multiwalled carbon nanotubes as adhesives to join graphite fibre reinforced polymer composites. *Nanotechnology*(7), 791.

Iijima. (1991). Helical microtubules of graphitic carbon. *Nature, 354*(6348), 56-58.

Irwin. (1957). Analysis of stresses and strains near the end of a crack traversing a plate. *J. Appl. Mech.* doi: citeulike-article-id:810675.

Iwahori, Ishikawa, Ishiwata, & Sumizawa. (2005). Mechanical properties improvements in two-phase and three-phase composites using carbon nano-fiber dispersed resin. *Composites Part A: Applied Science and Manufacturing, 36*(10 SPEC. ISS.), 1430-1439.

Joshi, & Dikshit. (2012). Enhancing interlaminar fracture characteristics of woven CFRP prepreg composites through CNT dispersion. [Article]. *Journal of Composite Materials, 46*(6), 665-675. doi: 10.1177/002199 8311410472.

Karapappas, Vavouliotis, Tsotra, Kostopoulos, & Paipetis. (2009). Enhanced fracture properties of carbon reinforced composites by the addition of multi-wall carbon nanotubes. *Journal of Composite*

Materials, 43 (Copyright 2009, The Institution of Engineering and Technology), 977-985.

Kepple, Sanborn, Lacasse, Gruenberg, & Ready. (2008). Improved fracture toughness of carbon fiber composite functionalized with multi walled carbon nanotubes. [Article]. *Carbon, 46*(15), 2026-2033. doi: 10.1016/j.carbon.2008.08.010.

Kim, & Mai. (1991). High-strength, high fracture-toughness fiber composites with interface control - a review. [Review]. *Composites Science and Technology, 41*(4), 333-378. doi: 10.1016/0266-3538(91)90072-w.

Li, Dichiara, Zha, Su, & Bai. (2014). On improvement of mechanical and thermo-mechanical properties of glass fabric/epoxy composites by incorporating CNT–Al2O3 hybrids. *Composites Science and Technology, 103*(0), 36-43. doi: http://dx.doi.org/10.1016/j.compscitech.2014.08.016.

Mathur, Chatterjee, & Singh. (2008). Growth of carbon nanotubes on carbon fibre substrates to produce hybrid/phenolic composites with improved mechanical properties. [Article]. *Composites Science and Technology, 68*(7-8), 1608-1615. doi: 10.1016/j.compscitech.2008.02.020.

Meyers, Chen, Lin, & Seki. (2008). Biological materials: Structure and mechanical properties. *Progress in Materials Science, 53*(1), 1-206. doi: http://dx.doi.org/10.1016/j.pmatsci.2007.05.002.

Miedlar, Berens, Gunderson, & Gallagher. (2006). *Damage tolerant design handbook: Guidelines for the analysis and design of damage tolerant aircraft structures*: United States Air Force (USAF).

Mirjalili, Ramachandramoorthy, & Hubert. (2014). Enhancement of fracture toughness of carbon fiber laminated composites using multi wall carbon nanotubes. *Carbon, 79*(0), 413-423. doi: http://dx.doi.org/10.1016/j.carbon.2014.07.084.

Qiu, Zhang, Wang, & Liang. (2007). Carbon nanotube integrated multifunctional multiscale composites. *Nanotechnology, 18*(27), -. doi: Artn 275708 Doi 10.1088/0957-4484/18/27/275708.

Romhany, & Szebenyi. (2009). Interlaminar crack propagation in MWCNT/fiber reinforced hybrid composites. [Article]. *Express Polymer Letters, 3*(3), 145-151. doi: 10.3144/expresspolymlett.2009.19.

Sadeghian, Gangireddy, Minaie, & Hsiao. (2006). Manufacturing carbon nanofibers toughened polyester/glass fiber composites using vacuum assisted resin transfer molding for enhancing the mode-I delamination resistance. *Composites Part A-Applied Science and Manufacturing, 37*(10), 1787-1795. doi: DOI 10.1016/j.compositesa.2005.09.010.

Sadeghian, Minaie, Gangireddy, & Hsiao. (2005). Mode-idelamination characterization for carbon nanofibers toughened polyester/glassfiber composites. Paper presented at the *50th International SAMPE Symposium and Exhibition*, May 1, 2005 - May 5, 2005, Long Beach, CA, United states.

Singh, Choudhary, Saini, & Mathur. (2012). Designing of epoxy composites reinforced with carbon nanotubes grown carbon fiber fabric for improved electromagnetic interference shielding. [Article]. *Aip Advances, 2*(2), 6. doi: 10.1063/1.4730043.

Tang, Ye, Zhang, & Deng. (2011). Characterization of transverse tensile, interlaminar shear and interlaminate fracture in CF/EP laminates with 10 wt% and 20 wt% silica nanoparticles in matrix resins. [Article]. *Composites Part A: Applied Science and Manufacturing, 42*(12), 1943-1950. doi: 10.1016/j.compositesa.2011.08.019.

Thostenson, Li, & Chou. (2005). Nanocomposites in context. *Composites Science and Technology, 65*(3-4), 491-516.

Veedu, Ghasemi-Nejhad, Kougen, Kar, Ajayan, Anyuan, Soldano, & Xuesong. (2006). Multifunctional composites using reinforced laminae with carbon-nanotube forests. *Nature materials, 5*(6), 457-462.

Wichmann, Sumfleth, Gojny, Quaresimin, Fiedler, & Schulte. (2006). Glass-fibre-reinforced composites with enhanced mechanical and electrical properties - Benefits and limitations of a nanoparticle modified matrix. *Engineering Fracture Mechanics, 73*(16), 2346-2359. doi: DOI 10.1016/j.engfracmech.2006.05.015.

Wicks, Wang, Williams, & Wardle. (2014). Multi-scale interlaminar fracture mechanisms in woven composite laminates reinforced with aligned carbon nanotubes. *Composites Science and Technology, 100*(0), 128-135. doi: http://dx.doi.org/10.1016/j.compscitech.2014.06.003.

Yokozeki, Iwahori, & Ishiwata. (2007). Matrix cracking behaviors in carbon fiber/epoxy laminates filled with cup-stacked carbon nanotubes (CSCNTs). *Composites Part A: Applied Science and Manufacturing, 38*(3), 917-924. doi: 10.1016/j.compositesa.2006.07.005.

Yokozeki, Iwahori, Ishiwata, & Enomoto. (2007). Mechanical properties of CFRP laminates manufactured from unidirectional prepregs using CSCNT-dispersed epoxy. *Composites Part A: Applied Science and Manufacturing, 38*(10), 2121-2130. doi: 10.1016/j.compositesa.2007.07.002.

Yuanxin, Jeelani, Pervin, & Lewis. (2008). Fabrication and characte-rization of carbon/epoxy composites mixed with multi-walled carbon nanotubes. *Materials science & engineering. A, Structural materials, 475*(1-2), 157-165.

Zhu, Imam, Crane, Lozano, Khabashesku, & Barrera. (2007). Processing a glass fiber reinforced vinyl ester composite with nanotube enhancement of interlaminar shear strength. *Composites Science and Technology, 67*(7-8), 1509-1517. doi: DOI 10.1016/j.compscitech.2006.07.018.

In: Advances in Aerospace Science ... ISBN: 978-1-53615-689-8
Editors: Parvathy Rajendran et al. © 2019 Nova Science Publishers, Inc.

Chapter 4

INFLUENCE OF AVIATION FUEL ON COMPOSITE MATERIALS

Sharmendran Kumarasamy[1], Nurul Musfirah Mazlan[1] and Aslina Anjang Ab Rahman[1,2]*

[1]School of Aerospace Engineering, Universiti Sains Malaysia,
Pulau Pinang, Malaysia
[2]Cluster for Polymer Composites,
Science and Engineering Research Centre, Universiti Sains Malaysia,
Pulau Pinang, Malaysia

ABSTRACT

Aviation jet fuel is the most consumed fuel type to power aircraft used in civil and military aviation sectors. Advancements in composite materials have recently been incorporated into the design of integral fuel tanks or 'wet wing' structures. The replacement of metallic materials in fuel tanks by composite materials has reduced aircraft weight and increased fuel storage space. This chapter discusses aviation jet fuel, its properties and the application of composite materials in aerospace. The effect of absorption and moisture diffusion behaviour on composite materials due to fuel uptake and water and seawater environments is also discussed. The

---

* Corresponding Author's Email: aeaslina@usm.my.

influences of moisture uptake, water, seawater and aviation fuel attack on composite materials are investigated. The effect of these solutions should be understood to ensure the compatibility and long-term use of composite materials, especially in the design of integral fuel tank structures. Guidelines and safety measures can be provided to the aerospace industry through such an understanding to limit or prevent failure caused by long-term exposure of composite materials to moisture and fuel attacks.

Keywords: composite material, mechanical property degradation, aviation fuel, diffusion, absorption

1. INTRODUCTION

A composite material is versatile and has various applications that range from a simple household item to a critical structural part. Structural tanks, such as storage and fuel tanks, are critical parts that have gained popularity, and they utilise composite materials in their application. In aircraft applications, integral fuel tanks made of composite materials have been developed to reduce aircraft weight and increase fuel storage space. The hot–wet event is a major issue in the aviation field. Degradation of mechanical properties exerts a significant effect under hot–wet and fully immersed conditions. The severe degradation of mechanical properties experienced by composite tanks due to fire risks may result in a catastrophic fire event. This chapter reviews the types of fuel used in the aviation industry and the effects of fuel uptake on the mechanical properties of composite materials. Considering that alternative renewable fuel types are being investigated for use in the aviation industry to reduce the consumption and dependence on petroleum-based fuel, renewable fuel alternatives, such as biodiesel and blended fuel composed of biodiesel and Jet-A, are discussed in this chapter. The moisture diffusion behaviour and effect of absorption on the performance of composite materials in water and seawater environments are evaluated to provide readers an overview of the topic.

2. BACKGROUND

Driving forces mainly depended on the speed and luxury of aircraft when the aviation industry began to boom. However, aircraft efficiency became the major driver in the modern age, and it has made the aviation industry what it is today. Aircraft systems are engineered annually by using cutting edge technologies to make them more aerodynamic and lighter than before and to increase their efficiency. Nevertheless, a major aspect has remained unchanged, that is, most flights still use the same fuel. Researchers are searching for alternative fuel that will eventually replace the current aviation fuel due to the depletion of fossil fuel. The next subtopic explains the types of fuel used in aerospace and the application of composite materials in aerospace.

2.1. Fuel in Aerospace

Fuel is the lifeblood of any vehicle, and the aviation industry is not exempted from this. High-quality fuel should be consistently supplied to aircraft to achieve optimum performance and efficiency. The most commonly used aviation fuel types are jet fuel and Avgas (Books et al., 2010). Jet fuel is used in jet engines where air is sucked into the turbine to produce a powerful thrust that propels the aircraft forward. Avgas, which stands for aviation gasoline, is normally used in turboprop engines on small aircraft, although it has started to become obsolete because most modern aircraft use gas turbines to generate thrust (Jetec, 2017). Different types of aviation fuel have specialised functions depending on aircraft applications.

2.2. Composites in Aerospace

A composite material is versatile and has various applications that range from household use to aerospace use. Composite materials helped propel the aerospace industry to what it is today. Engineers and designers can design and fabricate advanced structural parts by using composite materials due to the unique properties of these materials (Soutis, 2005). The use of composite materials dates to 1909 when the fuselage of a deHavilland Albatross transport aircraft was manufactured using a sandwich composite structure. Glass and carbon fibres were developed in the 1950s, and this development skyrocketed the aerospace industry to what is today (Potter, 1996). The typical fibres used in aerospace applications are shown in Table 4.

The two largest aircraft manufacturers, Boeing and Airbus, compete to increase the overall composite materials used in aircraft structures. As shown in Figure 8, the measure of composite materials utilised as a part of aeroplane assembly has shown a consistent increase from the 1970s (Ching Hao et al., 2014; Red et al., 2014; Slayton et al., 2015) because composite materials are lighter than metals but possess similar strength. This feature reduces the weight of the aircraft and increases its fuel efficiency and performance. In addition, large and complex parts of an aircraft can be manufactured using composite materials to reduce the total number of parts utilised. This usage also reduces the requirement for fasteners and joints on the aircraft.

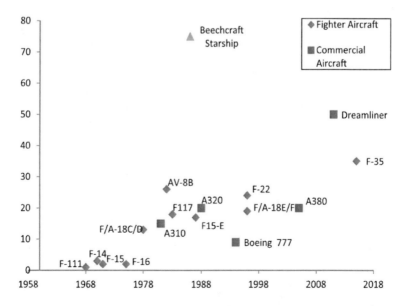

Figure 8. Percentage of composite materials used in aircraft (Slayton et al., 2015).

Table 4. Fibres used in aerospace applications (Nayak, 2014)

Fibres	Application
Glass	Aircraft interiors
	Aircraft secondary structures, radomes, fairing, rocker motor casings
	Small passenger aircraft
Aramid	Aircraft fairings
Carbon	Primary structures in aircraft and spacecraft
	widely used in many different parts, such as satellites, antenna dishes and aircraft

3. AVIATION FUEL PROPERTIES

Fuels can be partitioned into two classes, namely, chemical and nuclear. Chemical fuels can be additionally divided into solid, liquid and gaseous, and nuclear fuels can be further classified as fission or fusion depending on the reaction. Fuels are utilised to control everything around us by discharging heat energy, which is then converted to mechanical energy and to another form of energy, such as electric energy (Stark, 1966). The conversion is performed through a heat engine (Senft, 2007).

Several factors or properties may determine whether a fuel is suitable for a certain application. These properties are fuel flash point, kinematic viscosity, fuel density and calorific value. The temperature at which a fuel ignites when induced with a flame or spark is the fuel flash point (Ishida et al., 1986). Knowing the flash point is important for users to operate fuels within the temperature safety limit. The second property is fuel kinematic viscosity, which indicates the fuel's capability to flow (Moradi et al., 2012). This parameter determines the penetration period of gas droplets and its length during combustion. Knowing the density of a fuel determines its adequacy for engine combustion. This parameter directly affects the performance of the engine and its efficiency (Alptekin et al., 2008). Caloric value is the amount of energy, specifically heat energy, released during combustion for a specific amount of fuel (Palanna, 2009). This parameter indicates the amount of energy within a fuel.

3.1. Kerosene Fuel

Kerosene is derived from petroleum and generally utilised as a part of ventures and in households. Gas turbine engine aircraft are powered by kerosene. Aviation-grade kerosene is known as Jet-A fuel. A typical aircraft cruises at an altitude between 30,000 and 45,000 feet above sea level. At this altitude, the surrounding temperature ranges from −40°C to −57°C, but kerosene can maintain a low viscosity at this temperature. The physical properties of a typical Jet-A fuel are shown in Table 5. Various American Society of Testing and Materials (ASTM) methods have been applied to kerosene to meet the international standards for aviation fuel, and the results are shown in Table 3 (Speight, 2015; Xuan, 2016).

Table 5. Physical properties of Jet-A fuel (Chevron, 2004)

Properties	Jet-A
Flashpoint	38°C
Auto-ignition temperature	210°C
Freezing point	−40°C
Density at 15°C	0.820 kg/l
Specific energy	43.02 MJ/kg
Energy density	35.3 MJ/L

3.2. Biodiesel Fuel

ASTM defines biodiesel as a type of renewable fuel that encompasses monoalkyl esters of long fatty acid chains (Speight, 2015). These fatty acid chains normally originate from either vegetable or animal sources. The process of turning vegetable oil or animal fats into biodiesel is called transesterification (María Cerveró et al., 2008). This process relocates the glycerine in the molecule with low-molar-mass alcohol. This process results in a mixture that has physicochemical properties that are comparable with those of a petroleum diesel fuel. In the US, soybean oil is the primary wellspring of biodiesel, whereas biofuel based on rapeseed oil is the equivalent in Europe. Meanwhile, biodiesel in South East Asian countries is derived from canola, coconut, palm and corn oil (Hoekman et al., 2012).

Table 6. Jet-A fuel properties based on ASTM testing (Chevron, 2004; Xuan, 2016)

Properties		Jet-A	ASTM test method
Composition			
Acidity, total mg KOH/g	Max	0.1	D3242
Aromatics: one of the following requirements should be met:			
1. Aromatics, vol%	Max	25	D1319
2. Aromatics, vol%	Max	26.5	D6379
Sulfur, mercaptan, mass%	Max	0.003	D3227
Sulfur, total mass%	Max	0.3	D1266, D2622, D4294 or D5453
Volatility			
Distillation			D2887 or D86
Distillation temperature, °C			
10% recovered, temperature (T10)	Max	205	
50% recovered, temperature (T50)		Report	
90% recovered, temperature (T90)		Report	
Final boiling point, temperature	Max	300	
Distillation residue, %	Max	1.5	
Distillation loss, %	Max	1.5	
Flashpoint, °C	Min	38	D56 or D3828
Density at 15°C, kg/m^3		775–840	D1298 or D4052
Fluidity			
Freezing point, °C	Max	−40	D5972, D7153, D7154 or D2386
Viscosity −20°C, mm^2/s	Max	8	D445

Biodiesel has gained popularity as an environment-friendly alternative to traditional fossil fuel. Biodiesel is also renewable and biodegradable (Zhang et al., 2015). Manzanera et al. (2008) stated that a viable alternative fuel should have an environmental value, can be produced at a large scale economically without supply deficit, and can provide a net energy gain.

ASTM D6751 is used to determine whether or not the properties of biodiesel meet international standardised specifications (Bhuiya et al., 2016). The biodiesel properties based on ASTM D6751 that are used in this research are tabulated in Table 7. Additional properties of biodiesel fuel, such as cloud point, cetane number and pour point, are measured and calculated to obtain a thorough understanding. Cloud point is the temperature below which wax in diesel forms a cloudy appearance.

This temperature should be determined because it is an indicator that the oil has clogged the engine filter (Hammami et al., 2003). Meanwhile, pour point refers to the temperature below which the oil loses its flow mobility (Nikolaev et al., 2013). Oil with increased viscosity due to temperature decreases engine performance and efficiency. Cetane number is used as an indicator of the pressure required for the engine to ignite the biodiesel. This parameter refers to the speed at which biodiesel can combust (Sivaramakrishnan et al., 2012).

The properties of biodiesel may shift depending on the measure of unsaturated fat in its molecular structure. Before a biodiesel can be used in current engine systems, its properties should be similar or better than those of the currently used petroleum-based diesel fuel to ensure that currently available engines can be used without any modification. Gui et al. (2008) summarised and compared the physical and chemical properties of several types of biodiesel with those of a petroleum-derived diesel, and the data are tabulated in Table 5.

3.3. Kerosene and Biodiesel Blended Fuel

Petroleum is non-renewable fuel and has faced depletion in the past several years. By 2035, the worldwide interest in crude oil or petroleum is expected to increase by 30% (Sulaiman, 2007). The increase in the demand for a new aeroplane in the coming years has added to the expansion of worldwide petroleum demand (Airbus, 2011; Leahy, 2010; Nygren et al., 2009). According to the law of supply and demand, the overall cost of a product increases when demand increases and supply decreases. The price of crude oil or petroleum has increased in the past several years.

Currently, the main fuel supply for aircraft is kerosene, and kerosene is a by-product of petroleum, which can directly affect the aircraft overall cost. This condition has led to new studies on creating fuel blends that can reduce the usage of kerosene as the main fuel supply for aircraft. Biodiesel is a type of renewable fuel that is constantly available in the market because it is made from either vegetable oil or animal fats (Bello et al., 2000). Researchers and scientists have investigated kerosene and biodiesel blend as an alternative fuel for aeroplanes. This blend could aid in solving two main concerns, which are petroleum depletion and environmental pollution caused by burning fossil fuel (Sidjabat, 2013).

Table 7. Biodiesel properties (Bhuiya et al., 2016)

Property specification	Unit	Biodiesel ASTM D6751 Test Method	Limits
Flashpoint	°C	ASTM D93	130 min
Cloud point	°C	ASTM 2500	−3 to −12
Pour point	°C	ASTM 97	−15 to −16
Cloud filter pugging point	°C	ASTM D6371	Max +5
Cetane number		ASTM D613	47 min
Density at 15°C	kg/m^3	ASTM D1298	880
Kinematic viscosity at 40°C	mm^2/s	ASTN D445	1.9–6.0
Iodine number	gl$_2$/100 g	-	-
Acid number	mmKOH/g	ASTM D664	0.5 min
Oxidisation stability		-	-
Stoichiometric air/fuel ratio	w/w	ASTM PS121	13.8
Cold soak filtration	s	ASTM D6751	360
Carbon residue	% m/m	ASTM D4530	0.05 max
Copper corrosion		ASTM D130	No. 3 max
Distillation temperature	°C	ASTM D1160	360
Lubricity	M	ASTM D6079	520 max
Sulphated ash content	%mass	ASTM D874	0.002 max
Ash content	%mass	-	-
Water and sediment		ASTM D2709	0.005 vol% max

Table 8. Physical and chemical properties of different biodiesel types compared with those of petroleum-based diesel (Gui et al., 2008)

Parameter	Petroleum-based diesel	Soy-bean	Rapeseed	Palm	Castor	Rubber Seed
Viscosity at 40°C	2.6	4.08	4.5	4.42	-	5.81
Specific gravity	0.85	0.885	0.882	0.860-0.90	0.96	0.874
Calorific value (MJ/kg)	42	39.76	37	-	39.5	36.5
Flashpoint (°C)	68	69	170	182	260	130
Cloud point (°C)	-	-2	-4	15	-12	4
Pour point (°C)	-20	-3	-12	15	-32	-
Ash content (wt%)	0.01	-	-	0.02	0.02	-
Acid value (mg KOH/g)	-	-	-	0.08	-	0.118

Several test flights have been conducted on military aircraft and commercial airlines by using blends of Jet-A and biodiesel fuel (Fortier et al., 2014). The International Air Transport Association (IATA) has defined an objective to increase the usage of blended fuels to approximately 6% by 2020. Furthermore, greenhouse gas (GHG) emissions can be reduced by using blended fuel mixtures. Payan et al. (2014) compared the GHG emissions of several types of biodiesel fuel with those of kerosene and found that the GHG emission is much less than that of conventional aviation fuel. Table 9 shows a summary of GHG emissions of different biodiesel sources relative to kerosene.

Table 9. GHG emission reduction with different types of biodiesel (Payan et al., 2014)

Type of Biodiesel	GHG Emission Reduction from Kerosene Fuel (%)
Sweet sorghum	133
Camelina	86
Jatropha	42
Wood residues	148
Algae	124
Miscanthus	72
Switchgrass	63

The reduction of GHG emission in this type of fuel has led IATA to set a goal to reduce the aviation carbon footprint by 50% in 2050 (IATA, 2013). Figure 9 shows the IATA roadmap towards reducing GHG emissions by 2050. Four key strategies have been presented by IATA to reach this ambitious goal, and one of them is investing in a sustainable energy source, which is biofuel (IATA, 2009).

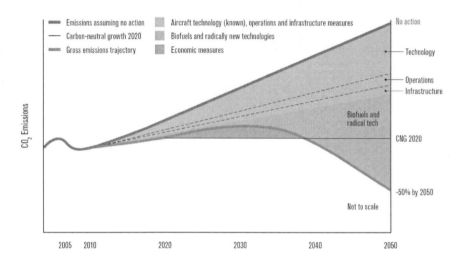

Figure 9. IATA's roadmap for the reduction of GHG emissions (IATA, 2013).

A study has been conducted on the compatibility of several blended biodiesels with Jet-A fuel. The results showed that blends that have approximately 10%–20% of methyl ester in their composition have similar properties as those of existing aviation fuel (Baroutian et al., 2013). The biodiesel used in this study was obtained from cheap feedstock, such as Jatropha Curcas, and this could reduce the fuel cost of an aircraft. Another study was performed by Dunn (2001) to investigate the practicability of using methyl soya ester (SME) as biodiesel in the fuel blend. The researcher found that before the blended fuel using SME can be used in a commercial aircraft cruising at 30,000 feet, additives must be utilised to decrease the fuel's freezing point because a pure SME blend starts to freeze at −29°C.

Depending on the esterification process using either ethanol or methanol, biodiesel production can have a long chain of either fatty acid ethyl esters or fatty acid methyl esters (FAMEs), respectively (Yusoff et al., 2014). A study has been conducted by taking a segment of this long chain, notably carbon chain 8 to 16, which has a similar order to Jet-A fuel, and by blending it with jet fuel (Llamas, Al-Lal, et al., 2012). The blended fuel meets several specifications, such as smoke point, density, flash point, freezing point and viscosity at −20°C of a conventional jet fuel based on the ASTM D1655 standard. The researcher concluded that blends that contain up to 10% of biodiesel based on volume is feasible to be used in commercial aircraft (Llamas, García-Martínez, et al., 2012).

Habib et al. (2010) investigated the engine performance of a small-scale gas turbine engine in terms of static thrust when mixed fuel is utilised. The overall static thrust of the 50% blend of biofuel had a slight decrease compared with pure Jet-A fuel. By contrast, the carbon emission by the blended fuel was lower than that of the Jet-A fuel. No alteration was required on the turbine engine to operate the blended fuel.

Abu Talib et al. (2014) used palm oil methyl ester biodiesel blended with Jet-A fuel to investigate the performance of an Armfield CM4 turbojet engine. Multiple blends based on biodiesel fuel were investigated, and the results showed that the 20% biodiesel blend has a comparable performance yield as conventional Jet-A fuel. Biofuel obtained from recycled canola cooking oil was tested on an SR-30 gas turbine engine (French, 2003). The fuel performance was slightly lower than that with Jet-A fuel but still acceptable.

4. MOISTURE DIFFUSION BEHAVIOUR OF COMPOSITE MATERIALS

Flory–Huggins theory states that a polymer can be regarded as a lattice of cells that can be filled with either a polymer molecule or a solvent molecule (Favre et al., 1996; Flory, 1953, 1970). Normally, the diffusion of liquid or gases into a material is modelled based on Fick's law of diffusion. Diffusion behaviour follows Fick's second law, which states that the concentration in a system changes over time and can be expressed as Eq. (1).

$$\frac{\partial M}{\partial t} = D \cdot \frac{\partial^2 M}{\partial Z^2}, \tag{1}$$

where M is the moisture content, D is the diffusion coefficient, t is the conditioning time and Z is the length in the thickness direction. As shown in Figure 10, a typical Fickian curve has two characteristics (M. Li, 2000). The first characteristic is the initial absorption phase, and the second one is the saturation phase. The initial phase has a linear absorption curve until a certain period where the absorption rate decreases and enters the second absorption phase. The second phase is where the net diffusion activity of the material is zero, which causes the curve to have a constant weight gain. A different type of diffusion curve that does not follow Fickian laws is called a non-Fickian sorption curve. This behaviour is normally due to other factors that interact with the Brownian motion of a normal Fickian curve (De Wilde et al., 1994; Lundgren et al., 1998).

Two main factors, namely, changes in the internal structure of the polymer material and its diffusion rate, should be considered when dealing with polymer material diffusion (J. A. Ferreira et al., 2015) because the polymer material is viscoelastic (Gross, 1953). A viscoelastic material is elastic and viscous when deformed. This property causes a delay reaction towards the Brownian motion, which is the principle in Fickian diffusion (J. Ferreira et al., 2014).

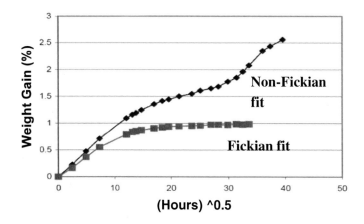

Figure 10. Resin weight gain against the square root of time (M. Li, 2000).

For a diffusion to obey Fick's law, the rate of diffusion must be either smaller or larger than the rate of relaxation of the material-fluid system (J. Ferreira et al., 2014). However, the solvent molecule that penetrates into the polymer material causes stress that results in an interference to the Brownian motion when the rate of diffusion is equal to the material relaxation rate; then, the diffusion curve becomes non-Fickian, as shown in Figure 11 (Thomas et al., 1980, 1981, 1982). Thomason (1995) investigated the water uptake curve of several resin systems. As shown in Figure 11, the moisture uptake curve shows a fairly Fickian behaviour although several of the resin systems require a long time to reach the saturation state.

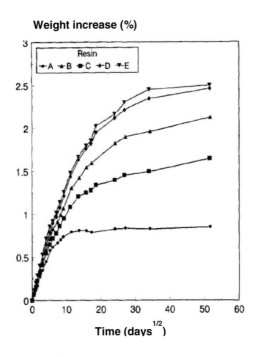

Figure 11. Moisture uptake curve for different types of resin (Thomason, 1995).

Dhakal et al. (2007) reported that several factors affect the diffusion curve of a material. Examples include the surrounding temperature, material hydrophilicity, void content and exposed surface area of the material to the solvent used. The formation of micro-voids, which propagate to micro-cracks, could also cause a delay in saturation because many solvent molecules could diffuse into the material (Alaneme et al., 2013). The formation of micro-voids and micro-cracks occurs due to the internal swelling of the material. The mechanical properties of the polymer material can also be permanently degraded by swelling (Bismarck et al., 2004).

As previously mentioned, the diffusion in a polymer-based material does not obey Fickian law in several cases due to the internal stress caused by the diffused moisture molecules (Neogi, 1983; Vrentas et al., 1998). This internal stress could cause damage to the polymer network, which causes a decrease in mass change, such as that observed by Kootsookos et al. (2004).

In addition, a Fickian model of diffusion cannot fully capture the sub-scale fluctuation due to the swelling and shrinking of the polymer system because the composite material is heterogeneous at the microscopic level (Alfrey et al., 1966; Dentz et al., 2006). Figure 12 shows the different types of non-Fickian sorption curve (De Kee et al., 2008; Neogi, 1996). Moisture content M_i at the initial stage can be presented as Eq. (2).

$$M_i = k \cdot t^n, \qquad (2)$$

where k and n are constants.

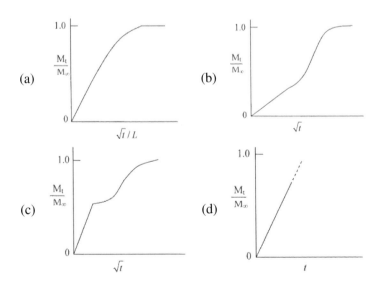

Figure 12. Different types of sorption curves: (a) classical Fickian, (b) sigmoidal, (c) two-step and (d) Case II (De Kee et al., 2008).

The value of n is equal to ½ for a classical Fickian curve, whereas the constant n is equal to 1 for a non-Fickian sorption curve or known as Case II sorption. Thus, the value of n for other anomalous sorptions can be expressed as ½ < n < 1 (ASTM, 1999). The swelling of the polymer material is related to the solubility parameter, as proposed by Hildebrand (Dixon-Garrett et al., 2000; Gerrard, 1976). The solubility factor, which affects the solubility of a material, can be divided into three components (Hansen, 2002). The components are dispersion force, hydrogen bonds and polar component. These components directly affect the behaviour of the polymer network when a solvent molecule penetrates the network.

Apart from Fickian behaviour sorption, another type of diffusion exists where the mass uptake shows a continuous but slow increment rather than reaching equilibrium (Davies et al., 2013). This process is called Langmuir's diffusion. Figure 13 shows a typical Langmuir's diffusion curve. This phenomenon normally occurs because of the unreacted epoxide group in a polymer network (Wong et al., 1985). This epoxide group undergoes a reversible hydrolysis reaction, which temporarily traps water molecules in the network.

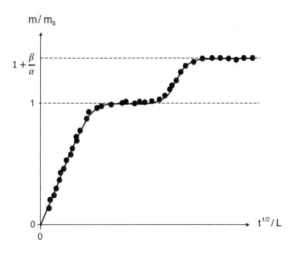

Figure 13. Langmuir's sorption curve (Davies et al., 2013).

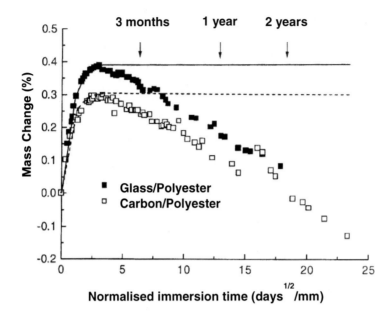

Figure 14. Seawater uptake curve for glass fibre polyester and carbon fibre polyester composites (Kootsookos et al., 2004).

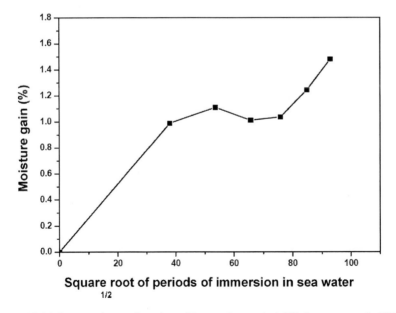

Figure 15. Moisture gain as a function of immersion period (Chakraverty et al., 2015).

The diffusion curve shown in Figure 7 starts out as a Fickian curve, but after being immersed for about 25 days, the solvent molecule begins to cause internal stress that disturbs the Fickian diffusion (Camino et al., 1998; Kootsookos et al., 2004; Prian et al., 1999). This process results in a non-Fickian diffusion behaviour.

Another researcher reported a diffusion behaviour similar to Langmuir's diffusion, as shown in Figure 15 (Chakraverty et al., 2015). The absorption curve may also be due to the formation of micro-cracks, which increases Langmuir's effects. A long immersion period can cause leaching of fibreglass, which increases the rate of penetration due to the availability of many empty spaces (Maxwell et al., 2005).

5. Effect of Absorption on the Performance of Composite Materials

The effect of liquid or moisture uptake on a composite is dependent on the individual segments, which are the resins, the fibre and the bonding between them (M. Hale et al., 1998). The effect of liquid/moisture uptake from several solutions, such as water, seawater and fuel, is discussed in the following sections.

5.1. Fuel Solutions

Composite materials are so versatile that their applications range from aerospace parts to typical household appliances. Composite materials are used as tanks, specifically fuel tanks. A fuel tank is a critical part of a vehicle because it holds the most flammable substance in a vehicle. Thus, its structural integrity of these tanks, especially aircraft fuel tanks, should remain unchanged for a long period to avoid fatal accidents. An accident on a plane causes a huge tragedy.

In this section, three types of aviation fuel solution, namely, Jet-A fuel, biodiesel fuel and blended fuel mixtures, are discussed. A researcher has investigated the effect of aviation fuel oil on composite materials (Ren et al., 2015). The results showed that the elasticity of composite specimens diminishes after treatment with such solutions. Figure 16 shows the signs of fibre and matrix debonding due to immersion in fuel.

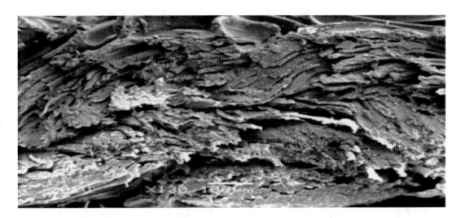

Figure 16. SEM image of the fracture surface of the composite after being exposed to jet fuel (Ren et al., 2015).

In another study, several glass fibre epoxy laminates were immersed in aviation fuel, and their mechanical properties were tested (Kumar et al., 2016). The average flexural stress of the immersed specimen decreased slightly compared with that of a standard specimen, as shown in Figure 17. Figure 18 shows the different failure mechanisms after the specimen underwent flexural testing. Compared with the original specimen, the immersed specimen experienced slight delamination in the area where the fibres broke (Ren et al., 2015).

Genanu (2011) performed a fatigue test on a glass fibre epoxy laminate after being immersed in a fuel solution. Micro-cracks formed in the specimens that were immersed, as shown in Figure 19. Moreover, the fracture region of the immersed specimens showed a slight brittle behaviour.

Another study was conducted on the effect of kerosene and diesel on steel–concrete composites (Fawzi et al., 2013). The compressive strength of the composite decreased after exposure to kerosene and diesel for 120 days. This result is mainly due to the weakening of the bond between the fibres and matrix (Abbas, 2017; Francis et al., 2010).

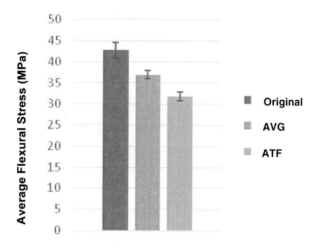

Figure 17. Average flexural stress of glass fibre under different conditioning (Kumar et al., 2016).

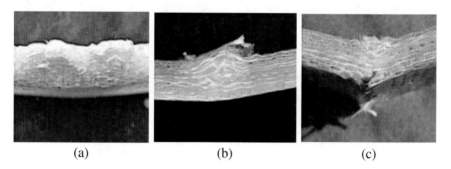

Figure 18. Flexural failure for (a) standard specimen, (b) AVG aviation fuel and (c) ATF aviation fuel (Kumar et al., 2016).

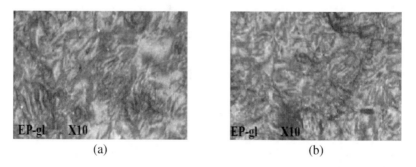

Figure 19. SEM images (a) before immersion of the specimen and (b) after immersion of the specimen (Genanu, 2011).

Meissner et al. (1944) performed a test on a mortar cube after conditioning in a high-octane environment and found that the general mechanical properties of the specimen do not affect its integrity because the decrement is insignificant. Similar to Meissner et al. (1944), Jasim et al. (2010) reached a similar conclusion where the decrement in compressive and tensile strengths is relatively small. In addition, they found that the decrement in mechanical properties is contrary to that of oil viscosity.

Schmidt (2010) stated that although biodiesel is not corrosive in the beginning, its molecular structure contains FAME, which can be hydrolysed by microbes. This reaction transforms the non-corrosive biodiesel into an organic acid with highly corrosive hydrogen sulfide (Nelson, 2010). The corrosive nature of hydrolysation could corrode steel tanks, which leads to oil leakage.

5.2. Water Environment

The diffusion of a water molecule into composites can affect its mechanical properties. Glass transition temperature (T_g) is the temperature where the material can change between glassy solid state and viscous liquid form (N. R. Jadhav et al., 2009a). The material is in liquid form when the temperature is above T_g, whereas the material remains in a solid glassy state when the temperature is below T_g (N. Jadhav et al., 2009b). The T_g of GFRP laminate is reduced due to the absorbed water acting as a plasticiser (Shen et al., 1965). This condition could cause a degradation in the mechanical properties of GFRP (Kelley Frank et al., 2003). A similar behaviour was reported using carbon fibre composites (Hughes, 1991; Selzer et al., 1997; Wyatt et al., 1969). Figure 20 shows that the T_g of a material decreases with the increase in moisture uptake. The T_g of the specimen decreases from approximately 200°C to 125°C with approximately 6% moisture uptake. The good thing about this change is that it is reversible and can be recovered by drying the material (Schwartz, 1997).

Water uptake for a short period can increase mode I fracture toughness, but prolonged exposure can cause the toughness of GFRP to worsen

(Marom, 1989). The diffused water molecule degrades the interfacial bond between the fibre reinforcement and resin matrix (Schultheisz et al., 1997). A weak bond between the fibre and matrix causes the GFRP composite to experience delamination easily (Assarar et al., 2011). Zhong et al. (2015) reported that moisture uptake can cause a slight reduction in the adhesive between fibres and matrixes. Figure 21 shows that a specimen loses its interface strength after being subjected to a water environment (Hu et al., 2016). The interface strengths in the beginning of the experiment in [95] were 43 and 35 MPa. The interface strengths decreased to 27 and 29 MPa after exposure to distilled water for approximately 350 days. The specimen's interface strength was evaluated by using a fibre push-out test.

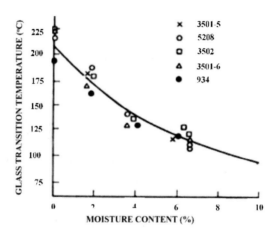

Figure 20. Glass transition temperature versus moisture uptake (Schwartz, 1997).

Another study was conducted by Chakraverty et al. (2017) on the effect of humidity and temperature on the diffusivity rate of a water molecule into a GFRP composite. The inter-laminar shear strength (ILSS) diminished because of the increased moisture taken up by the glass fibre epoxy laminate. Figure 15 shows the degradation in the ILSS of a glass fibre epoxy laminate as a function of conditioning period. The ILSS of the specimen decreases by approximately 24% and 18% for hygrothermal and hydrothermal environments, respectively. A prolonged immersion period causes permanent degradation, such as the formation of micro-cracks, which induce increased moisture uptake and result in a high plasticiser effect on the polymer (Abdel-Magid et al., 2005; Delasi et al., 1978; Hahn et al., 1978). In another study, a disc-shaped micro-crack was observed at a size of 40 μm by using an optical microscope, as shown in Figure 23 (Gautier et al., 2000).

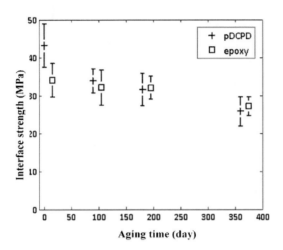

Figure 21. Interface strength after immersion in distilled water (Hu et al., 2016).

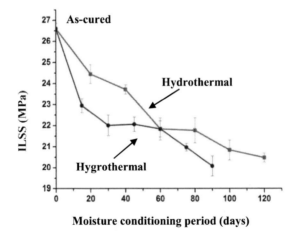

Figure 22. ILSS of glass fibre epoxy laminate versus moisture uptake period (Chakraverty et al., 2017).

Figure 23. Disc-shaped crack formed in the matrix (size 40 μm) (Gautier et al., 2000).

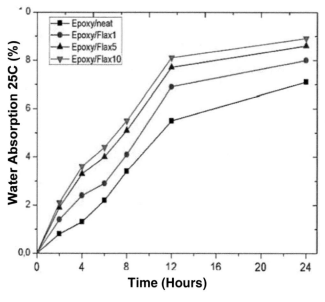

Figure 24. Water absorption rate at 25°C (Huner, 2015).

Huner (2015) investigated the water absorption in a flax fibre-reinforced epoxy composite and found that the hydroxide group in the flax fibre simulates the diffusivity of the water molecule into the composite. Figure 24 shows the moisture uptake of the flax-reinforced epoxy composite at 25°C. The specimens start to reach the saturation state after immersion for approximately 12 hours. The specimen with the highest flax concentration has the highest moisture uptake curve at approximately 0.8%, whereas the specimen without flax has the lowest moisture uptake curve at 0.5%. Increasing the flax weight in the composite causes increased water uptake because an increased number of hydroxide groups are present in the composite. In addition, the voids in the laminates act as a microchannel for the water molecule to diffuse through capillary action (Guermazi et al., 2014; Narendar et al., 2013).

Another study found that the addition of fibre to epoxy can reduce the water uptake by increasing the tortuosity path for the water to diffuse (Chow, 2007). As shown in Figure 25, the uptake percentage is reduced from approximately 2.4% to approximately 1.1% in saturation. The absorption reaches quasi-equilibrium before reaching equilibrium due to the formation of micro-voids (Popineau et al., 2005). In addition, the water molecule in the epoxy matrix can reduce the modulus of the composite according to the findings of Jiang et al. (2016).

A study was performed by Mohanty et al. (2016) to explore the water absorption behaviour and its consequences on the flexural strength of GFRP composites. He discovered that moisture uptake detrimentally affects the flexural quality of the glass fibre composite, as shown in Figure 26. The glass fibre epoxy composite showed a decrease of about 19% in flexural strength. Meanwhile, the flexural strength of the glass carbon hybrid epoxy composite was reduced by 34%. The longer the immersion period of the composite specimens was, the higher the detrimental effect on their mechanical properties was.

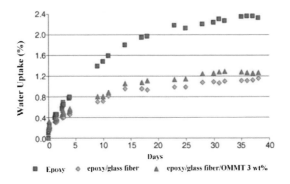

Figure 25. Moisture uptake of epoxy, epoxy/glass fibre composite and epoxy/glass fibre/OMMT nanocomposites versus time (Chow, 2007).

A water molecule that has diffused into the polymer network forces itself to occupy the spaces between the networks (Lassila et al., 2002). Water molecules acting as a plasticiser cause the polymer network to move and increase its mobility, which results in a decrease in flexural strength (Anusavice, 2003). Another researcher observed a similar pattern where the flexural strength and modulus decrease with the increase in the exposure period of the specimen to a water environment (Bal et al., 2015). This condition is due to the plasticiser effect by water molecules on the polymer.

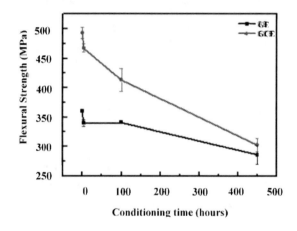

Figure 26. Flexural strength of glass/epoxy and glass/carbon hybrid epoxy composites versus moisture uptake (Mohanty et al., 2016).

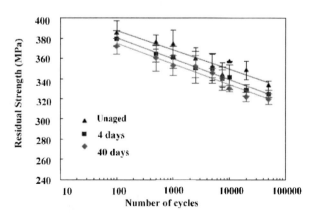

Figure 27. Residual strength of different GFRP conditions based on the number of cycles (Y et al., 2015).

The decrease in the specimen's stiffness and strength, as shown in Figure 20 and Figure 21, is due to the damaged fibre–matrix interface of the composite by water molecules (Y et al., 2015). This condition indicates that the composite material experiences mechanical damage due to water diffusion, as mentioned by Schultheisz et al. (1997).

Silva et al. (2016) directly evaluated the impacts of ecological conditions on the molecular structure of epoxy. After immersing a specimen in a water environment, the specimen components were observed using a scanning electron microscope and compared with those of a standard specimen as a reference. Figure 29 shows the comparison of the immersed specimen and a virgin specimen. The chemical composition remained constant after a long immersion period.

In addition, the specimen that was submerged in a water environment experienced some form of permanent degradation, as shown in Figure 30. The arrows show the interfacial cracks between the fibres and fibre/matrix debonding (Imielińska et al., 2004). This condition increased the diffusivity rate of the water molecule into the specimen because many spaces that can be filled up were created in the matrix (Komada et al., 2017; Shirrell, 1978).

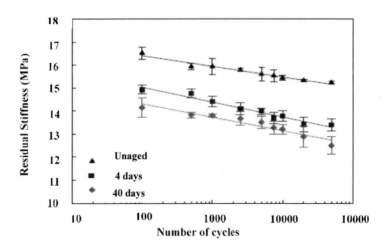

Figure 28. Residual stiffness of different GFRP conditions based on the number of cycles (Y et al., 2015).

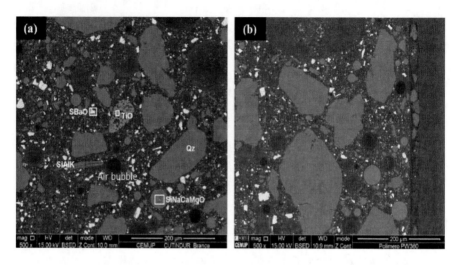

Figure 29. Epoxy constitution under SEM: (a) reference specimen and (b) specimen after immersion in distilled water (Silva et al., 2016).

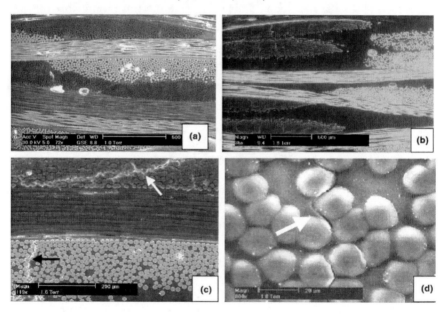

Figure 30. SEM image of the aramid–glass hybrid epoxy composite specimen: (a) standard, (b–d) immersed with degradation form (Imielińska et al., 2004).

5.3. Seawater Environment

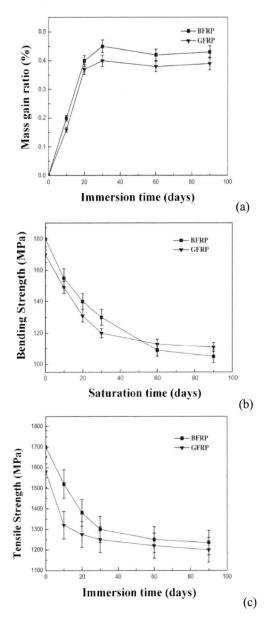

Figure 31. Moisture uptake correlation of glass fibre and basalt fibre epoxy composite: (a) sorption curve, (b) flexural stress and (c) tensile stress (Wei et al., 2011).

Table 10. Composition of seawater (Mourad et al., 2010)

Element	Parts per million
Chloride (Cl)	21700
Sodium (Na)	19497
Sulfate (SO$_4$)	2880
Magnesium (Mg)	1938
Potassium (K)	743
Calcium (Ca)	602
Bicarbonate (HCO$_3$)	200

Exposure to a seawater environment can also cause degradation in composite materials (Deniz et al., 2012; Grant et al., 1995; Hasnine, 2010). The major composition of seawater is shown in Table 10 (Mourad et al., 2010). Seawater absorption could result in damage, such as formation of micro-cracks, propagation of voids and swelling (Barkoula et al., 2009; Cabanelas et al., 2003; Garcia-Espinel et al., 2015; Gu, 2009; Visco et al., 2011). A study was conducted on the impact of seawater on the mechanical properties of GFRP composites, and the results showed that the elastic and flexural stresses decrease with the increase in immersion period (Wei et al., 2011), as shown in Figure 31.

Another researcher monitored the effect of seawater on the degradation of GFRP composites (Wang et al., 2015). As shown in Figure 32, the adhesive between the fibres and matrix degrade with the increase in immersion period (Feng et al., 2014; Wang et al., 2015) because the hydroxyl group responds to the moisture particle, which results in the debonding between the fibres and matrix (Heshmati et al., 2016). As shown in Figure 33, the tensile strength of the glass fibre composite worsens as a function of immersion time (Wang et al., 2015). After immersion for six months, the tensile strength of the specimen decreases from approximately 450 MPa to 390 MPa. This condition is due to the weakening of the interface between the fibres and matrix (Aldajah et al., 2009; Bal et al., 2015).

Influence of Aviation Fuel on Composite Materials 103

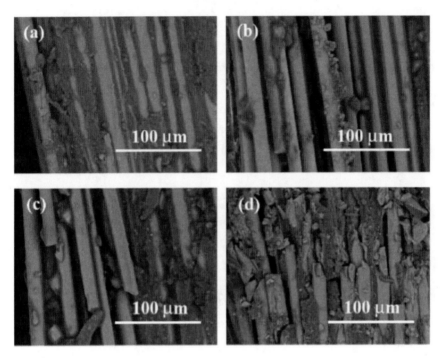

Figure 32. GFRP laminate under SEM: (a) standard specimen, (b) 2 months of seawater immersion, (c) 4 months of seawater immersion and (d) 6 months of seawater immersion (Wang et al., 2015).

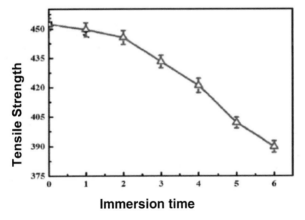

Figure 33. Tensile strength of GFRP laminate versus immersion period in seawater (Wang et al., 2015).

The salts in seawater penetrate into the glass fibre with different depths over time, and this affects the moisture uptake of the specimen (Chakraverty et al., 2015). Energy dispersive spectroscopy (EDS) is used to define the penetration depth of salt elements into the matrix. Figure 34 and Figure 35 show the penetration for 2 and 12 months of immersion period, respectively. The bulky attributes of the salts advance the augmentation of the polymer network, resulting in a decrease in the T_g of the specimens (Diamant et al., 1981; Gellert et al., 1999).

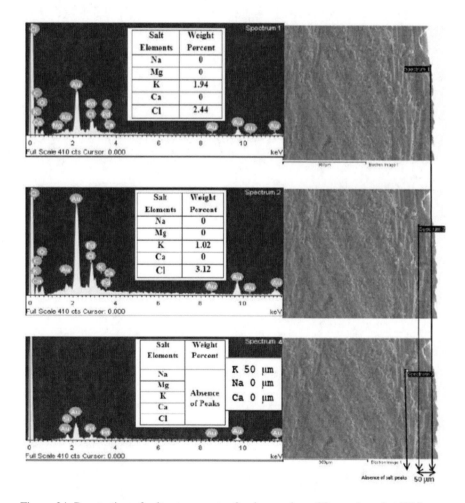

Figure 34. Penetration of salt components after immersion of 2 months using EDS spectra (Chakraverty et al., 2015).

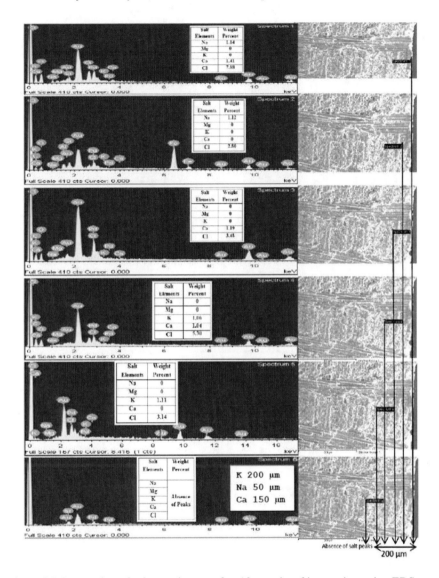

Figure 35. Penetration of salt constituents after 12 months of immersion using EDS spectra (Chakraverty et al., 2015).

Another researcher immersed a GFRP bar, which is primarily used for strengthening in civil buildings, in an alkaline solution (Micelli et al., 2017). The tensile strength slightly decreased, and the decrement is insignificant. This condition is essential due to the interaction of the alkaline ions and the silicon ion from the fibre (Karbhari et al., 2003). This reaction damages the fibre's structure. Crystal formation in the interstices between the fibres could cause localised damage (Chin Joannie et al., 2004; Chin Joannie et al., 1999; Maljaee et al., 2016).

As shown in Figure 36, the diffusion of seawater into the specimen decreases with the tensile strength of the glass fibre (Abdel-Magid et al., 2005; Wang et al., 2015). This result is due to the plasticiser effect and the weakened interfacial bonding (Ghiassi et al., 2013; Sciolti et al., 2010). Figure 37 shows the seawater damage on the glass fibre of the GFRP composite. The interfacial bond between the fibre and matrix decreases as a function of exposure period (Daniel et al., 2006; Xu et al., 2012).

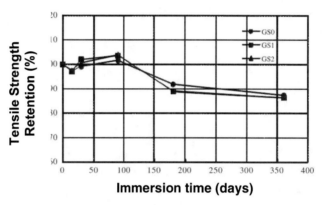

Figure 36. Tensile strength of the glass fibre specimen versus immersion period (Wang et al., 2015).

Figure 37. Optical microscopic images of (a) standard dry specimen and (b) specimen exposed to seawater (Xu et al., 2012).

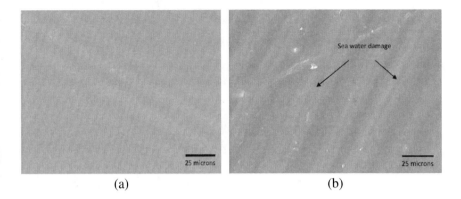

Figure 38. SEM images of (a) dry specimen and (b) specimen immersed in seawater (Xu et al., 2012).

Damage by seawater occurs on the outer ply of the fibre, and the damage is not expected to reach the inner ply upon saturation (G. Li et al., 2004; Xu et al., 2012). As shown in Figure 38, the glass fibre exposed to the seawater environment exhibits interfacial damage (Krishnan, 2010; Xu et al., 2012). Signs of fibre and matrix debonding are observed.

Feng et al. (2014) conducted an experiment on the corrosive nature of seawater towards glass fibre composites. He found that the resin on the surface was dissolved due to the alkaline nature of seawater. The degradation increased with the increase in the pH value (Stamenović et al., 2011). Aldajah et al. (2009) obtained a similar result on the mechanical properties of GFRP laminates when conditioning in seawater. The mechanical properties of the specimen decreased with long immersion period in a seawater environment (Maljaee et al., 2016).

CONCLUSION

According to literature, several factors, such as the type of resin used and void content of the composite, affect the diffusion of a liquid or solvent into the polymer network of the composite. The resin used should be able to withstand corrosion from the solvent used. The epoxy resin has better mechanical properties than other thermoset resins, such as polyester. In addition, the manufacturing process, which directly affects the void content of the composite, has a high influence on the diffusion rate of the solution molecule during immersion. This process can cause the composite to exhibit non-Fickian diffusion.

Although the current knowledge on the diffusion of fuel solutions into composite materials is limited, we can conclude that composites experience slight degradation due to fuel attack. Furthermore, the usage of blended fuel is in demand, especially in the aviation industry, and the sorption curve reported by previous researchers may not represent it accurately due to the variation in blend ratios.

Some of the reported degradations on composite materials due to immersion are from the formation of micro-voids and micro-cracks, weakening of the interlaminar force between the fibre and matrix and leaching of low-molecular-weight materials. Some of the reported degradations are permanent and affect the composite mechanical properties.

Another crucial finding from the literature survey is that most previous researchers only considered the mechanical properties of composites at room temperature, and limited research has been performed at high temperatures and under fire. In addition, the effect of post-fire mechanical properties on a composite immersed in fuel solution has not been investigated. Research data on the effect of high temperature, fire properties and post-fire properties due to fuel uptake or absorption in a composite material fuel tank should be obtained to develop a detailed understanding of the issue.

ACKNOWLEDGMENT

This study was funded by Universiti Sains Malaysia Bridging Grants 304/PAERO/6316087 and 304/PAERO/6316194.

REFERENCES

Abbas. (2017). Effect of Kerosene and Gas Oil Products on Different Types of Concrete. *International Journal of Science and Research (IJSR)*, 6(12), 1718-1722. doi: 10.21275/art20178987.

Abdel-Magid, Ziaee, Gass, & Schneider. (2005). The combined effects of load, moisture and temperature on the properties of E-glass/epoxy composites. *Composite Structures, 71*(3), 320-326. doi: https://doi.org/10.1016/j.compstruct.2005.09.022.

Abu Talib, Gires, & Ahmad. (2014). Performance Evaluation of a Small-Scale Turbojet Engine Running on Palm Oil Biodiesel Blends. *Journal of Fuels, 2014*, 9. doi: 10.1155/2014/946485.

Airbus. (2011). *Global Market Forecast 2011-2030: Airbus.*

Alaneme, Oke, & A Omotoyinbo. (2013). *Water absorption characteristics of polyester matrix composites reinforced with oil palm ash and oil palm fibre* (Vol. 2).

Aldajah, Alawsi, & Rahmaan. (2009). Impact of sea and tap water exposure on the durability of GFRP laminates. *Materials & Design, 30*(5), 1835-1840. doi: https://doi.org/10.1016/j.matdes.2008.07.044.

Alfrey, Gurnee, & Lloyd. (1966). Diffusion in glassy polymers. *Journal of Polymer Science Part C: Polymer Symposia, 12*(1), 249-261. doi: 10.1002/polc.5070120119.

Anusavice. (2003). *Phillips' Science of Dental Materials - eBook*: Elsevier Health Sciences.

Assarar, Scida, El Mahi, Poilâne, & Ayad. (2011). Influence of water ageing on mechanical properties and damage events of two reinforced composite materials: Flax–fibres and glass–fibres. *Materials & Design, 32*(2), 788-795. doi: https://doi.org/10.1016/j.matdes.2010.07.024.

ASTM. (1999). Test Method F739-99a Standard Test Method for Resistance of Protective Clothing Materials to Permeation by Liquids or Gases Under Conditions of Continuous Contact. *ASTM International.* doi: 10.1520/f0739-99a.

Bal, & Saha. (2015). Effect of sea and distilled water conditioning on the overall mechanical properties of carbon nanotube/epoxy composites. *International Journal of Damage Mechanics, 26*(5), 758-770. doi: 10.1177/1056789515615184.

Barkoula, Paipetis, Matikas, Vavouliotis, Karapappas, & Kostopoulos. (2009). *Environmental degradation of carbon nanotube-modified composite laminates: A study of electrical resistivity* (Vol. 45).

Baroutian, Aroua, Raman, Shafie, Ismail, & Hamdan. (2013). Blended aviation biofuel from esterified Jatropha curcas and waste vegetable oils. *Journal of the Taiwan Institute of Chemical Engineers, 44*(6), 911-916. doi: https://doi.org/10.1016/j.jtice.2013.02.007.

Bello, Torres, Herrera, & Sarmiento. (2000). The Effect of Diesel Properties on the Emissions of Particulate Matter. *CT&F - Ciencia, Tecnología y Futuro, 2*, 31-46.

Bhuiya, Rasul, Khan, Ashwath, Azad, & Hazrat. (2016). Prospects of 2nd generation biodiesel as a sustainable fuel – Part 2: Properties, performance and emission characteristics. *Renewable and Sustainable Energy Reviews, 55*, 1129-1146. doi: https://doi.org/10.1016/j.rser.2015.09.086.

Bismarck, Aranberri-Askargorta, Springer, Lampke, Wielage, Stamboulis, Shenderovich, & Limbach. (2004). Surface characterization of flax, hemp and cellulose fibers; Surface properties and the water uptake behavior. *Polymer Composites, 23*(5), 872-894. doi: 10.1002/pc.10485

Books, & Wikipedia. (2010). *Aviation Fuels: Kerosene, Avgas, Jet Fuel, Aviation Fuel, Zip Fuel, Aircraft Fuel System, Microbial Corrosion, Jp-8*: General Books.

Cabanelas, Prolongo, Serrano, Bravo, & Baselga. (2003). Water absorption in polyaminosiloxane-epoxy thermosetting polymers. *Journal of Materials Processing Technology, 143-144*, 311-315. doi: https://doi.org/10.1016/S0924-0136(03)00480-1.

Camino, Polishchuk, Luda, Revellino, Blancon, & Martinez-Vega. (1998). Water ageing of SMC composite materials: a tool for material characterisation. *Polymer Degradation and Stability, 61*(1), 53-63. doi: https://doi.org/10.1016/S0141-3910(97)00129-8.

Chakraverty, Mohanty, Mishra, & Biswal. (2017). Effect of Hydrothermal immersion and Hygrothermal Conditioning on Mechanical Properties of GRE Composite. *IOP Conference Series: Materials Science and Engineering, 178*, 012013. doi: 10.1088/1757-899x/178/1/012013.

Chakraverty, Mohanty, Mishra, & Satapathy. (2015). Sea Water Ageing of GFRP Composites and the Dissolved salts. *IOP Conference Series: Materials Science and Engineering, 75*(1), 012029.

Chevron. (2004). *Aviation Fuels Technical Review*.
Chin Joannie, Aouadi, Haight Michael, Hughes William, & Nguyen. (2004). Effects of water, salt solution and simulated concrete pore solution on the properties of composite matrix resins used in civil engineering applications. *Polymer Composites, 22*(2), 282-297. doi: 10.1002/pc.10538.
Chin Joannie, Nguyen, & Aouadi. (1999). Sorption and diffusion of water, salt water, and concrete pore solution in composite matrices. *Journal of Applied Polymer Science, 71*(3), 483-492. doi: 10.1002/(sici)1097-4628(19990118)71:3<483::aid-app15>3.0.co;2-s.
Ching Hao, Sapuan, & Hassan. (2014). *A Review of the Flammability Factors of Kenaf and Allied Fibre Reinforced Polymer Composites* (Vol. 2014).
Chow. (2007). Water absorption of epoxy/glass fiber/organo-montmorillonite nanocomposites. *Express Polymer Letters, 1*(2), 104-108. doi: 10.3144/expresspolymlett.2007.18.
Daniel, & Ishai. (2006). *Engineering Mechanics of Composite Materials* (Second ed.). New York: Oxford University Press.
Davies, & Rajapakse. (2013). *Durability of Composites in a Marine Environment*: Springer Netherlands.
De Kee, Liu, & Hinestroza. (2008). Viscoelastic (Non-Fickian) Diffusion. *The Canadian Journal of Chemical Engineering, 83*(6), 913-929. doi: 10.1002/cjce.5450830601.
De Wilde, & Frolkovic. (1994). The modelling of moisture absorption in epoxies: effects at the boundaries. *Composites, 25*(2), 119-127. doi: https://doi.org/10.1016/0010-4361(94)90005-1.
Delasi, & Whiteside. (1978). *Effect of Moisture on Epoxy Resins and Composites*. 2-2-19. doi: 10.1520/stp34855s.
Deniz, & Karakuzu. (2012). Seawater effect on impact behavior of glass–epoxy composite pipes. *Composites Part B: Engineering, 43*(3), 1130-1138. doi: https://doi.org/10.1016/j.compositesb.2011.11.006.
Dentz, & Tartakovsky Daniel. (2006). Delay mechanisms of non-Fickian transport in heterogeneous media. *Geophysical Research Letters, 33*(16). doi: 10.1029/2006gl027054.

Dhakal, Zhang, & Richardson. (2007). Effect of water absorption on the mechanical properties of hemp fibre reinforced unsaturated polyester composites. *Composites Science and Technology, 67*(7), 1674-1683. doi: https://doi.org/10.1016/j.compscitech.2006.06.019.

Diamant, Marom, & Broutman. (1981). The effect of network structure on moisture absorption of epoxy resins. *Journal of Applied Polymer Science, 26*(9), 3015-3025. doi: 10.1002/app.1981.070260917.

Dixon-Garrett, Nagai, & Freeman. (2000). Sorption, diffusion, and permeation of ethylbenzene in poly(1-trimethylsilyl-1-propyne). *Journal of Polymer Science Part B: Polymer Physics, 38*(8), 1078-1089. doi: 10.1002/(sici)1099-0488(20000415)38:8<1078::aid-polb8>3.0.co;2-2.

Dunn. (2001). Alternative Jet Fuels from Vegetable Oils. *Transactions of the ASAE, 44*(6), 1751. doi: https://doi.org/10.13031/2013.6988.

Favre, Nguyen, Clement, & Neel. (1996). Application of Flory-Huggins theory to ternary polymer-solvents equilibria: A case study. *European Polymer Journal, 32*(3), 303-309. doi: https://doi.org/10.1016/0014-3057(95)00146-8.

Fawzi, & AL-Ameer. (2013). Effect of Petroleum Products on Steel Fiber Reinforced Concrete. *Journal of Engineering, 19*(1).

Feng, Wang, Wang, Loughery, & Niu. (2014). Effects of corrosive environments on properties of pultruded GFRP plates. *Composites Part B: Engineering, 67*, 427-433. doi: https://doi.org/10.1016/j.compositesb.2014.08.021.

Ferreira, de Oliveira, da Silva, & Simon. (2014). Molecular Transport in Viscoelastic Materials: Mechanistic Properties and Chemical Affinities. *SIAM Journal on Applied Mathematics, 74*(5), 1598-1614. doi: 10.1137/140954027.

Ferreira, Grassi, Gudiño, & de Oliveira. (2015). A new look to non-Fickian diffusion. *Applied Mathematical Modelling, 39*(1), 194-204. doi: https://doi.org/10.1016/j.apm.2014.05.030.

Flory. (1953). *Principles of Polymer Chemistry* (1st ed.). Ithaca, United States: Cornell University Press.

Flory. (1970). Fifteenth Spiers Memorial Lecture. Thermodynamics of polymer solutions. *Discussions of the Faraday Society, 49*, 7. doi: 10.1039/df9704900007.

Fortier, Roberts, Stagg-Williams, & Sturm. (2014). Life cycle assessment of bio-jet fuel from hydrothermal liquefaction of microalgae. *Applied Energy, 122*, 73-82. doi: https://doi.org/10.1016/j.apenergy.2014.01.077.

Francis, A., Matitoli, Smith, & S. (2010). Relining of Potable Water Tanks for Strength and Corrosion Resistance *Corrosion and preventation* (pp. 1-9).

French. (2003). Recycled fuel performance in the SR-30 gas turbine. *Proceedings of the 2003 American Society for Engineering Education Annual Conference and Exposition,* Session 1133.

Garcia-Espinel, Castro-Fresno, Parbole Gayo, & Ballester-Muñoz. (2015). Effects of sea water environment on glass fiber reinforced plastic materials used for marine civil engineering constructions. *Materials & Design (1980-2015), 66*, 46-50. doi: https://doi.org/10.1016/j.matdes.2014.10.032.

Gautier, Mortaigne, Bellenger, & Verdu. (2000). Osmotic cracking nucleation in hydrothermal-aged polyester matrix. *Polymer, 41*(7), 2481-2490. doi: https://doi.org/10.1016/S0032-3861(99)00383-3.

Gellert, & Turley. (1999). Seawater immersion ageing of glass-fibre reinforced polymer laminates for marine applications. *Composites Part A: Applied Science and Manufacturing, 30*(11), 1259-1265. doi: https://doi.org/10.1016/S1359-835X(99)00037-8.

Genanu. (2011). *Study the Effect of Immersion in Gasoline and Kerosene on Fatigue Behavior for Epoxy Composites Reinforcement with Glass Fiber.*

Gerrard. (1976). Hildebrand's Solubility Parameters. In Gerrard (Ed.), *Solubility of Gases and Liquids: A Graphic Approach Data — Causes — Prediction* (pp. 55-57). Boston, MA: Springer US.

Ghiassi, Marcari, Oliveira, & Lourenço. (2013). Water degrading effects on the bond behavior in FRP-strengthened masonry. *Composites Part B:*

Engineering, 54, 11-19. doi: https://doi.org/10.1016/j.compositesb. 2013.04.074.

Grant, & Bradley. (1995). In-Situ Observations in SEM of Degradation of Graphite/Epoxy Composite Materials due to Seawater Immersion. *Journal of Composite Materials, 29*(7), 852-867. doi: 10.1177/ 002199839502900701.

Gross. (1953). *Mathematical Structure of the Theories of Viscoelasticity* Hermann & Cie. Paris, France.

Gu. (2009). Behaviours of glass fibre/unsaturated polyester composites under seawater environment. *Materials & Design, 30*(4), 1337-1340. doi: https://doi.org/10.1016/j.matdes.2008.06.020.

Guermazi, Haddar, Elleuch, & Ayedi. (2014). Investigations on the fabrication and the characterization of glass/epoxy, carbon/epoxy and hybrid composites used in the reinforcement and the repair of aeronautic structures. *Materials & Design (1980-2015), 56*, 714-724. doi: https://doi.org/10.1016/j.matdes.2013.11.043.

Gui, Lee, & Bhatia. (2008). Feasibility of edible oil vs. non-edible oil vs. waste edible oil as biodiesel feedstock. *Energy, 33*(11), 1646-1653. doi: https://doi.org/10.1016/j.energy.2008.06.002.

Habib, Parthasarathy, & Gollahalli. (2010). Performance and emission characteristics of biofuel in a small-scale gas turbine engine. *Applied Energy, 87*(5), 1701-1709. doi: https://doi.org/10.1016/j.apenergy. 2009.10.024.

Hahn, & Kim. (1978). *Swelling of Composite Laminates.* 98-98-23. doi: 10.1520/stp34860s.

Hammami, Ratulowski, & Coutinho. (2003). *Cloud Points: Can We Measure or Model Them?* (Vol. 21).

Hansen. (2002). *Hansen Solubility Parameters: A User's Handbook*: CRC Press.

Hasnine. (2010). *Durability of carbon fiber/vinylester composites subjected to marine environments and electrochemical interactions.* Master of Science, Florida Atlantic University.

Heshmati, Haghani, & Al-Emrani. (2016). Effects of moisture on the long-term performance of adhesively bonded FRP/steel joints used in bridges.

Composites Part B: Engineering, 92, 447-462. doi: https://doi.org/10.1016/j.compositesb.2016.02.021.

Hoekman, Broch, Robbins, Ceniceros, & Natarajan. (2012). Review of biodiesel composition, properties, and specifications. *Renewable and Sustainable Energy Reviews*, 16(1), 143-169. doi: https://doi.org/10.1016/j.rser.2011.07.143.

Hu, Li, Lang, Zhang, & Nutt. (2016). Water immersion aging of polydicyclopentadiene resin and glass fiber composites. *Polymer Degradation and Stability*, 124, 35-42. doi: https://doi.org/10.1016/j.polymdegradstab.2015.12.008.

Hughes. (1991). The carbon fibre/epoxy interface—A review. *Composites Science and Technology*, 41(1), 13-45. doi: https://doi.org/10.1016/0266-3538(91)90050-Y.

Huner. (2015). Effect of Water Absorption on the Mechanical Properties of Flax Fiber Reinforced Epoxy Composites. *Advances in Science and Technology Research Journal*, 9, 1-6. doi: 10.12913/22998624/2357.

IATA. (2009). *A global approach to reducing aviation emissions*, Montreal: IATA.

IATA. (2013). *The IATA Technology Roadmap Report* (4 ed.).

Imielińska, & Guillaumat. (2004). The effect of water immersion ageing on low-velocity impact behaviour of woven aramid–glass fibre/epoxy composites. *Composites Science and Technology*, 64(13), 2271-2278. doi: https://doi.org/10.1016/j.compscitech.2004.03.002.

Jadhav, Gaikwad, Nair, & Kadam. (2009a). Glass transition temperature: Basics and application in pharmaceutical sector. *Asian Journal of Pharmaceutics*, 3(2), 82. doi: 10.4103/0973-8398.55043.

Jadhav, Gaikwad, Nair, & Kadam. (2009b). *Glass transition temperature: Basics and application in pharmaceutical sector.*

Jasim, & Jawad. (2010). Effect of Oil on Strength of Normal and High Performance Concrete. *Al-Qadisiya Journal For Engineering Sciences*, 3(1).

Jetec. (2017). *Focus on Fuel Part One: Different Types of Aviation Fuel.*

Jiang, Song, Qiang, Kolstein, & Bijlaard. (2016). Moisture Absorption/Desorption Effects on Flexural Property of Glass-Fiber-Reinforced

Polyester Laminates: Three-Point Bending Test and Coupled Hygro-Mechanical Finite Element Analysis. *Polymers, 8*(8). doi: 10.3390/polym8080290.

Karbhari, Chin, Hunston, Benmokrane, Juska, Morgan, Lesko, Sorathia, & Reynaud. (2003). Durability Gap Analysis for Fiber-Reinforced Polymer Composites in Civil Infrastructure. *Journal of Composites for Construction, 7*(3), 238-247. doi: 10.1061/(asce)1090-0268(2003)7: 3(238).

Kelley Frank, & Bueche. (2003). Viscosity and glass temperature relations for polymer-diluent systems. *Journal of Polymer Science, 50*(154), 549-556. doi: 10.1002/pol.1961.1205015421.

Komada, Inagaki, Ueda, Omori, Hosaka, Tagami, & Miura. (2017). Influence of water immersion on the mechanical properties of fiber posts. *Journal of Prosthodontic Research, 61*(1), 73-80. doi: https://doi.org/10.1016/j.jpor.2016.05.005.

Kootsookos, & Mouritz. (2004). Seawater durability of glass- and carbon-polymer composites. *Composites Science and Technology, 64*(10), 1503-1511. doi: https://doi.org/10.1016/j.compscitech.2003.10.019.

Krishnan. (2010). *The Interfacial Failure of Bonded Materials and Composites.* PhD, Vanderbilt University, Nashville, Tennessee.

Kumar, & Lohchab. (2016). Influence of Aviation Fuel on Mechanical properties of Glass Fiber-Reinforced Plastic Composite. *International Advanced Research Journal in Science, Engineering and Technology, 3*(4). doi: 10.17148/IARJSET.2016.3413.

Lassila, Nohrström, & Vallittu. (2002). The influence of short-term water storage on the flexural properties of unidirectional glass fiber-reinforced composites. *Biomaterials, 23*(10), 2221-2229. doi: https://doi.org/10.10 16/S0142-9612(01)00355-6.

Leahy. (2010). *Airbus Global Market Forecast 2010-2029:* Airbus.

Li. (2000). *Temperature and Moisture Effects on Composite Materials For Wind Turbine Blades.* Master of Science in Chemical Engineering, Montana State University.

Li, Pang, Helms Jack, & Ibekwe Samuel. (2004). Low velocity impact response of GFRP laminates subjected to cycling moistures. *Polymer Composites, 21*(5), 686-695. doi: 10.1002/pc.10222.

Llamas, Al-Lal, Hernandez, Lapuerta, & Canoira. (2012). *Biokerosene from Babassu and Camelina Oils: Production and Properties of Their Blends with Fossil Kerosene* (Vol. 26).

Llamas, García-Martínez, Al-Lal, Canoira, & Lapuerta. (2012). Biokerosene from coconut and palm kernel oils: Production and properties of their blends with fossil kerosene. *Fuel, 102*, 483-490. doi: https://doi.org/10.1016/j.fuel.2012.06.108.

Lundgren, & Gudmundson. (1998). A Model for Moisture Absorption in Cross-Ply Composite Laminates with Matrix Cracks. *Journal of Composite Materials, 32*(24), 2226-2253. doi: 10.1177/00219983 9803202403.

M. Hale, & Gibson. (1998). *Coupon Tests of Fibre Reinforced Plastics at Elevated Temperatures in Offshore Processing Environments* (Vol. 32).

Maljaee, Ghiassi, Lourenço, & Oliveira. (2016). Moisture-induced degradation of interfacial bond in FRP-strengthened masonry. *Composites Part B: Engineering, 87*, 47-58. doi: https://doi.org/10.1016/j.compositesb.2015.10.022.

Manzanera, Molina-Muñoz, & Gonzalez-Lopez. (2008). *Biodiesel: An Alternative Fuel* (Vol. 2).

María Cerveró, Coca Prados, & Luque. (2008). *Production of biodiesel from vegetable oils* (Vol. 59).

Marom. (1989). Chapter 10 - Environmental Effects on Fracture Mechanical Properties of Polymer Composites. In Friedrich (Ed.), *Composite Materials Series* (Vol. 6, pp. 397-424): Elsevier.

Maxwell, S, Broughton, R, Dean, D, Sims, & D. (2005). *Review of accelerated ageing methods and lifetime prediction techniques for polymeric materials*.

Meissner, & Pearson. (1944). Tests of Gasoline-Resistant Coatings. *ACI Proc, 15*(6), 292. doi: 10.14359/8660.

Micelli, Corradi, Aiello, & Borri. (2017). Properties of Aged GFRP Reinforcement Grids Related to Fatigue Life and Alkaline Environment. *Applied Sciences, 7*(9). doi: 10.3390/app7090897.

Mohanty, Singh, Mahato, Rathore, Prusty, & Ray. (2016). Water absorption behavior and residual strength assessment of glass/epoxy and glass-carbon/epoxy hybrid composite. *IOP Conference Series: Materials Science and Engineering, 115*, 012029. doi: 10.1088/1757-899x/115/1/012029.

Mourad, Abdel-Magid, El-Maaddawy, & Grami. (2010). Effect of Seawater and Warm Environment on Glass/Epoxy and Glass/Polyurethane Composites. *Applied Composite Materials, 17*(5), 557-573. doi: 10.1007/s10443-010-9143-1.

Narendar, & Dasan. (2013). *Effect of Chemical Treatment on the Mechanical and Water Absorption Properties of Coir Pith/Nylon/ Epoxy Sandwich Composites* (Vol. 18).

Nayak. (2014). Composite Materials in Aerospace Applications *International Journal of Scientific and Research Publications (IJSRP), 4*(9).

Nelson. (2010). *Biodiesel corrosion could cause leaks in fuel infrastructure,* from https://www.mnn.com/earth-matters/energy/stories/biodieselcorrosion-could-cause-leaks-in-fuel-infrastructure.

Neogi. (1983). Anomalous diffusion of vapors through solid polymers. Part II: Anomalous sorption. *AIChE Journal, 29*(5), 833-839. doi: 10.1002/aic.690290519.

Neogi. (1996). *Diffusion in polymers*. New York: Marcel Dekker.

Nikolaev, V. Egorov, Nikolaev, & Sultanova. (2013). *A Method of Testing the Pour Point of Petroleum Products on Refrigerated Sloping Surface//Petrol.Sci.Technol.,* 2013, v. 31, N3, pp. 276-283 (Vol. 31).

Nygren, Aleklett, & Höök. (2009). Aviation fuel and future oil production scenarios. *Energy Policy, 37*(10), 4003-4010. doi: https://doi.org/10.1016/j.enpol.2009.04.048.

Payan, Kirby, Justin, & Mavris. (2014). *Meeting Emissions Reduction Targets: A Probabilistic Lifecycle Assessment of the Production of Alternative Jet Fuels.*

Popineau, Rondeau-Mouro, Sulpice-Gaillet, & Shanahan. (2005). Free/bound water absorption in an epoxy adhesive. *Polymer, 46*(24), 10733-10740. doi: https://doi.org/10.1016/j.polymer.2005.09.008.

Potter. (1996). *An Introduction to Composite Products: Design, Development and Manufacture.* Springer, 5th Ed.

Prian, & Barkatt. (1999). Degradation mechanism of fiber-reinforced plastics and its implications to prediction of long-term behavior. *Journal of Materials Science, 34*(16), 3977-3989. doi: 10.1023/a:1004647511910.

Red, & C. (2014). *Composite World,* from http://www.compositesworld.com/blog/2014.

Ren, Zhu, Men, Zhang, Guo, Yang, & Liu. (2015). *The effect of oil fouling on the mechanical and tribological properties of nomex fabric/phenolic composite* (Vol. 50).

Schmidt. (2010). Microbes Quickly Degrade a Popular Biofuel. *Chemical & Engineering News,* 10052110113143. doi: 10.1021/cen052110113143.

Schultheisz, McDonough, Kondagunta, Schutte, Macturk, McAuliffe, & Hunston. (1997). *Effect of moisture on E-glass/epoxy interfacial and fiber strengths* (Vol. 1242).

Schwartz. (1997). *Composite materials.* Upper Saddle River, N.J.: Prentice Hall PTR.

Sciolti, Frigione, & Aiello. (2010). *Wet Lay-Up Manufactured FRPs for Concrete and Masonry Repair: Influence of Water on the Properties of Composites and on Their Epoxy Components* (Vol. 14).

Selzer, & Friedrich. (1997). Mechanical properties and failure behaviour of carbon fibre-reinforced polymer composites under the influence of moisture. *Composites Part A: Applied Science and Manufacturing, 28*(6), 595-604. doi: https://doi.org/10.1016/S1359-835X(96)00154-6.

Shen, & Tobolsky. (1965). *Glass Transition Temperature of Polymers. 48,* 27-34. doi: 10.1021/ba-1965-0048.ch002.

Shirrell. (1978). *Diffusion of Water Vapor in Graphite/Epoxy Composites.* 21-21-22. doi: 10.1520/stp34856s.

Sidjabat. (2013). *The Characteristics of a Mixture of Kerosene and Biodiesel as a Substituted Diesel Fuel.*

Silva, Fernandes, Sena-Cruz, Xavier, Castro, Soares, & Carneiro. (2016). Effects of different environmental conditions on the mechanical characteristics of a structural epoxy. *Composites Part B: Engineering, 88*, 55-63. doi: https://doi.org/10.1016/j.compositesb.2015.10.036.

Sivaramakrishnan, & Ravikumar. (2012). *Determination of cetane number of biodiesel and it's influence on physical properties* (Vol. 7).

Slayton, & Spinardi. (2015). *Radical innovation in scaling up: Boeing's Dreamliner and the challenge of socio-technical transitions* (Vol. 47).

Soutis. (2005). Fibre reinforced composites in aircraft construction. *Progress in Aerospace Sciences, 41*(2), 143-151. doi: https://doi.org/10.1016/j.paerosci.2005.02.004.

Speight. (2015). *Handbook of Petroleum Product Analysis*: Wiley.

Stamenović, Putić, Rakin, Medjo, & Čikara. (2011). Effect of alkaline and acidic solutions on the tensile properties of glass–polyester pipes. *Materials & Design, 32*(4), 2456-2461. doi: https://doi.org/10.1016/j.matdes.2010.11.023.

Sulaiman. (2007). *Production of biodiesel: Possibilities and challenges* (Vol. 1).

Thomas, & Windle. (1980). A deformation model for Case II diffusion. *Polymer, 21*(6), 613-619. doi: https://doi.org/10.1016/0032-3861(80)90316-X.

Thomas, & Windle. (1981). Diffusion mechanics of the system PMMA-methanol. *Polymer, 22*(5), 627-639. doi: https://doi.org/10.1016/0032-3861(81)90352-9.

Thomas, & Windle. (1982). A theory of case II diffusion. *Polymer, 23*(4), 529-542. doi: https://doi.org/10.1016/0032-3861(82)90093-3.

Thomason. (1995). The interface region in glass fibre-reinforced epoxy resin composites: 2. Water absorption, voids and the interface. *Composites, 26*(7), 477-485. doi: https://doi.org/10.1016/0010-4361(95)96805-G.

Visco, Campo, & Cianciafara. (2011). Comparison of seawater absorption properties of thermoset resins based composites. *Composites Part A: Applied Science and Manufacturing, 42*(2), 123-130. doi: https://doi.org/10.1016/j.compositesa.2010.10.009.

Vrentas, Vrentas, & Huang. (1998). Anticipation of anomalous effects in differential sorption experiments. *Journal of Applied Polymer Science, 64*(10), 2007-2013. doi: 10.1002/(sici)1097-4628(19970606)64: 10<2007::aid-app15>3.0.co;2-3.

Wang, GangaRao, Liang, Zhou, Liu, & Fang. (2015). Durability of glass fiber-reinforced polymer composites under the combined effects of moisture and sustained loads. *Journal of Reinforced Plastics and Composites, 34*(21), 1739-1754. doi: 10.1177/0731684415596846.

Wei, Cao, & Song. (2011). Degradation of basalt fibre and glass fibre/epoxy resin composites in seawater. *Corrosion Science, 53*(1), 426-431. doi: https://doi.org/10.1016/j.corsci.2010.09.053.

Wong, & Broutman. (1985). Moisture diffusion in epoxy resins Part I. Non-Fickian sorption processes. *Polymer Engineering & Science, 25*(9), 521-528. doi: 10.1002/pen.760250903.

Wyatt, & Ashbee. (1969). Debonding in carbon fibre/polyester resin composites exposed to water: Comparison with 'E' glass fibre composites. *Fibre Science and Technology, 2*(1), 29-40. doi: https://doi.org/10.1016/0015-0568(69)90029-3.

Xu, Krishnan, Ning, & Vaidya. (2012). A seawater tank approach to evaluate the dynamic failure and durability of E-glass/vinyl ester marine composites. *Composites Part B: Engineering, 43*(5), 2480-2486. doi: https://doi.org/10.1016/j.compositesb.2011.08.039.

Xuan. (2016). *Study on the Effect of Biodiesel/Kerosene Mixture on Flame Propagation, Flame Temperature Distribution and Aircraft Engine Performance.* Bachelor of Engineering, Universiti Sains Malaysia.

Y, Mahi, & M. (2015). Effect of Fatigue Testing and Aquatic Environment on the Tensile Properties of Glass and Kevlar Fibers Reinforced Epoxy Composites. *Journal of Aeronautics & Aerospace Engineering, 04*(03). doi: 10.4172/2168-9792.1000150.

Yusoff, Xu, & Guo. (2014). Comparison of Fatty Acid Methyl and Ethyl Esters as Biodiesel Base Stock: a Review on Processing and Production Requirements. *Journal of the American Oil Chemists' Society, 91*(4), 525-531. doi: 10.1007/s11746-014-2443-0.

Zhang, Zhou, Chang, Pan, Liu, Li, Hu, & Yang. (2015). Production and fuel properties of biodiesel from Firmiana platanifolia L.f. as a potential non-food oil source. *Industrial Crops and Products, 76*, 768-771. doi: https://doi.org/10.1016/j.indcrop.2015.08.002.

Zhong, & Joshi. (2015). Impact behavior and damage characteristics of hygrothermally conditioned carbon epoxy composite laminates. *Materials & Design (1980-2015), 65*, 254-264. doi: https://doi.org/10.1016/j.matdes.2014.09.030.

In: Advances in Aerospace Science ...   ISBN: 978-1-53615-689-8
Editors: Parvathy Rajendran et al.   © 2019 Nova Science Publishers, Inc.

Chapter 5

DETERIORATION IN AERO-ENGINES

Koh Wei Chong, Senebahvan Maniam,
Aslina Anjang Ab Rahman and Nurul Musfirah Mazlan[*]
School of Aerospace Engineering, Universiti Sains Malaysia,
Pulau Pinang, Malaysia

ABSTRACT

Aero-engine is a mechanical turbomachinery that is subjected to substantial wear and tear throughout its service life. Efficiency of engine components can be achieved by increasing the power and decreasing the fuel consumption of the engine. Efficiency may vary depending on the exhaust gas temperature. Over the past few years, engine design has undergone several innovations and improvements to achieve general engine efficiency. The exhaust gas temperature is increased to even 1600 K to increase engine power. However, such an increase seriously affects engine components. For example, an increase in heat stress influences turbine blades. Moreover, exposure to high temperatures, particularly in the combustion chamber, may result in thermal distress due to combustion instabilities, thus degrading the efficiency of components. Many factors can lead to such deterioration, which consequently affects the performance of the engine and the lifespan of engine components. This chapter clarifies

---

[*] Corresponding Author's Email: nmusfirah@usm.my.

this topic by providing detailed factors that influence engine deterioration, the typical mechanisms of engine deterioration, examples of deterioration in engine components and the effect of engine deterioration on engine performance and engine component lifespan.

Keywords: deterioration, aero-engine, factors, mechanisms, engine performance

1. INTRODUCTION

Mechanical turbomachinery, such as gas turbine engines, are subjected to substantial wear and tear throughout their service life. Gas turbines undergo gradual deterioration of engine performance as the operation time increases (Döring, Staudacher, Koch, & Weißschuh, 2017). In industrial gas turbines, 70%–85% of overall performance deterioration is estimated to be caused by deposition (Diakunchak, 1992). Large particles that cause erosion can be removed from the fluid through appropriate filtration. However, the remaining large fraction of small particles that cause deposition is difficult to remove (Rainer Kurz & Brun, 2012). Unlike in industrial gas turbines, the deterioration of components in aero-engine gas turbines is classified as 'critical' or 'non-critical'. The classification depends on the effects of malfunction on aircraft integrity.

Deterioration in industrial gas turbines can be classified into three categories, namely, recoverable, non-recoverable and permanent (Diakunchak, 1992). However, in aircraft, deterioration can be categorised as (1) on-wing recoverable performance deterioration, (2) off-wing recoverable performance deterioration and (3) permanent performance deterioration. On-wing recoverable performance deterioration can be addressed through on-wing maintenance, such as compressor washing. The occurrence of a single or multiple random events, such as material failures and foreign object damage (FOD), requires operational procedures during take-off, engine operational habits, airport regulations and maintenance procedures to preserve the engine condition (Venediger, 2013). Off-wing recoverable performance deterioration can be addressed by off-wing maintenance, such as disassembly and replacement or refurbishment of damaged parts. Permanent performance deterioration cannot be recovered at economically justifiable expenses. This deterioration usually occurs due to natural ageing, which is an unavoidable process.

Deposition and erosion in aircraft engines are unique because they superimpose each other. Therefore, quantification of individual effects from in-service data is challenging. Evaluation using Pratt & Whitney JT9D and General Electric's CF06 turbofan engine in-service data has shown that one of the main factors causing performance deterioration of high-pressure compressors is surface finish degradation (Sallee, 1978). Among compressor components, front stages are more affected than rear stages, and stators are more affected than rotors; the deposition of mass on stators is approximately twice as high as that on rotors (Tarabrin, Schurovsky, Bodrov, & Stalder, 1998). Rapid roughness is estimated to build up within the first 1,000 flight cycles and remains approximately constant afterwards. Severe performance deterioration due to erosion and deposition has been reported by aircraft engine operators (Richardson, Sallee, & Smakula, 1979; Sallee, 1978; Ziemianski & Mehalic, 1980). Engines are damaged frequently during taxiing, take-off and landing by objects ejected from the ground by the landing gear or thrust reverser. Moreover, the inlet-vortex (or ground-vortex) phenomenon causes FOD.

A continuous airworthiness program has been developed to monitor and assess the condition of engines continuously and to ensure reliable and safe gas turbine operation, thus alleviating engine damage due to engine component degradation. The instrument called engine degradation monitoring or engine health monitoring (EHM) is essential for gas turbine operation because it determines engine performance and allows a prognosis for future performance trend prediction and estimation (Roth, Doel, & Cissell, 2005). Typically, the minimum recorded data for performance analysis consist of the pressure and temperature of the engine gas path, shaft rotational speed and fuel flow. In addition to these parameters, vibration data, engine oil temperature and pressure are used for such analysis. Apart from obtaining these parameters through sensors, visual inspection is performed either by using direct visualisation tools, such as a borescope, or indirect visual methods, such as non-destructive testing (Li, 2002).

A common measure primarily used to determine actual engine condition in terms of operational performance is the measurement of the engine exhaust gas temperature (EGT), which in turn allows for the calculation of the EGT margin (EGTM). This method is used by many major airline operators and has been proven to provide reliable results on engine health. It is usually measured at locations after the high-pressure turbine exit or in the first stages of the low-pressure turbine.

2. AERO-ENGINE GAS TURBINE

Most of the major propulsion systems for civil aircraft in service today use gas turbine engines, with turbofan engines being the most widely used engine variant for short-to-medium and long-range applications. One of the advantages of this engine type is that turbofan engines offer relatively good fuel efficiency and low noise levels compared with other turbojet engines. In addition, turbofan aero-engines are designed to work with prominent quality and efficiency under a wide range of operating conditions. They play a pivotal role in the flight capabilities of modern aircraft (Ruichao, Yingqing, Nguang, & Yifeng, 2018). In terms of flight velocity, turbofan engines have an efficient operation Mach number of up to 0.85, which is an advantage compared with a turboprop engine (Grieb, 2004). With the available technology at present, current turbofan engines have a two- or three-spool design ranging from medium to high bypass ratios. A typical turbofan engine consists of a low-pressure compressor (LPC), high-pressure compressor (HPC), combustor chamber (also known as a burner), high-pressure turbine (HPT), low-pressure turbine (LPT) and convergent core nozzle through the core of the engine (hot flow). Cold flow (bypass air) also runs through the fan and the convergent fan nozzle, as shown in Figure 39.

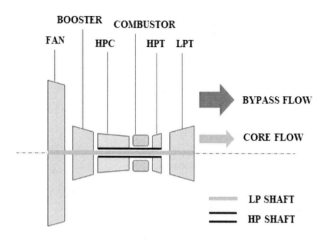

Figure 39. Typical two-shaft turbofan engine configuration (Venediger, 2013).

Several key parameters define engine performance to meet a given design specification. The two key parameters that describe the performance of an aircraft gas turbine engine are net thrust (FN) and specific fuel consumption (SFC) or thrust specific fuel consumption (TSFC). SFC is generally influenced by thermal, propulsive and combustion efficiency (Walsh & Fletcher, 2004). A turbofan engine also has three main design parameters, which are turbine entry temperature (TET), overall pressure ratio (OPR) and bypass ratio (BPR). A change in these three parameters significantly affects the engine's thermal and propulsive performance.

The maximum TET in aero-engine combustors is generally limited by the mechanical integrity of the combustion chamber and turbine parts, which are exposed to the highest gas temperatures in the entire engine. Apart from using available materials for manufacturing, these highly stressed engine parts can be applied with active cooling to ensure efficient operation. Hence, an engine that allows high TET normally exhibits good thermal performance (Bräunling, 2015). OPR represents the relationship of the total pressure at the compressor exit with the total pressure at the engine inlet. It heavily depends on the number of compressors and individual compressor design (i.e., number of stages). The maximum overall pressure ratios in aero-engines are generally limited by the maximum permissible engine weight and the operation ranges of the combustor and turbines.

Engine BPR is the ratio between the air mass flow rate of air that bypasses the core of the engine to the air mass flow rate passing through the core engine. The air that passes through the core is involved in the combustion process. The maximum engine BPR in aero-engines is usually limited by the size of the fan diameter or the decrease in the size of the core engine diameter. An extremely large fan increases the aircraft's total drag disproportionally and the weight of the fan section. Moreover, a large fan requires high shaft speed. Meanwhile, decreasing the size of the core engine is limited by compressor stage pressure ratio and the size of the combustion chamber. A more detailed elaboration of the correlations amongst engine OPR, BPR and fan pressure ratio with detailed analyses and diagrams can be found in the work of Walsh, P.P. and P. Fletcher (Walsh & Fletcher, 2004) and Bräunling, W.J. (Bräunling, 2015).

3. FACTORS THAT AFFECT ENGINE DEGRADATION

Many factors can affect engine deterioration, and these include environmental condition, materials of the engine components and changes in aircraft role.

3.1. Environmental Condition

An aircraft operates in a wide range of environmental conditions that change throughout the mission. For example, commercial aircraft required to travel in different climatic conditions at the start and end of their journey. Military aircraft also travel in different operating environments depending on their speed, manoeuvrability and impact of the battle space. Changes in an aircraft's operating condition, such as temperature, affect various parts of an aircraft engine in different ways. For example, on cold days, lubricants lose their viscosity, thus creating friction and wear issues for moving parts. Low temperatures may cause plastics and rubber parts to become brittle.

Metals also contract at different rates and change their tolerance at low temperatures.

3.2. Material Selection for Gas Turbine Components

Aero-engine components are subjected to a combination of mechanical cyclic and thermal loading, which is also known as thermo-mechanical fatigue, due to high-temperature and high-pressure regions, thereby causing fatigue and creep. This combination can seriously influence the failure mechanisms induced within the material with which the components are made. Adverse operational conditions or manufacturing defects, such as using insufficiently durable types of heat-resistant protective coatings or inappropriately applying blade materials, cause blade damage. Therefore, careful selection of protective coatings and materials for blade manufacture must consider mechanical and thermal properties. Aside from the material used in the turbine blade, blade design plays important roles in increasing durability. The shape of the turbine blade must be designed carefully to avoid vibration resonance during engine interruption. In addition, dampers are often used to eliminate dangerous forms and vibration frequency (Hyde, Sun, & Hyde, 2011).

3.3. Changes in Aircraft Role

Changes in aircraft role can influence engine component degradation. Scenarios, such as take-off incursion and rudder/tail failure, require the aircraft to change its role, leading to changes in engine operation and engine life.

Take-off incursion is a condition wherein the lift-off distance is suddenly decreased. This situation may occur when an object suddenly enters the runaway. Therefore, the aircraft has to take-off on an extremely short runway by increasing the thrust beyond the engine's maximum nominal thrust level. This additional thrust produced by the engine accelerates the aircraft and enables it to reach its take-off speed at a reduced distance. The engine is penalised due to increases in turbine temperature and rotor speeds beyond the current limits. Although the additional thrust may require only a few minutes, the over-thrust function may be utilised for long periods to overcome drag.

Rudder/tail failure occurs when the aircraft experiences either airframe of internal damage due to loss of hydraulics. This situation reduces the effectivity of flight control surfaces. The rudder or vertical stabiliser controls the yaw angle. Therefore, engine responsivity is necessary to control the yaw angle during rudder/tail failure. Engine responsivity can be obtained through increased accelerations that lead to a decrease the HPC surge margin.

4. Typical Mechanisms of Engine Deterioration

Engine degradation begins from the alteration of several mechanical and/or chemical properties of the gas turbine engine parts. The degradation of aerodynamic components, such as the engine compressor, combustor and turbine, under harsh environments is a major cause of engine performance deterioration. Deterioration in aircraft engines can be categorised as follows:

- On-wing recoverable performance deterioration
- Off-wing recoverable performance deterioration
- Permanent performance deterioration

Mechanical wear, thermal distress, abrasion, corrosion and erosion are typical mechanisms in engine degradation that cause the original shape, properties and condition of parts to change (Ranier Kurz & Brun, 2007). A general overview of these deterioration mechanisms and their effects is presented in this section.

4.1. Mechanical Wear

Mechanical wear often occurs in engine oil and oil seals in all parts of the engine. It causes an increase in leakage flow over time. Moving parts, such as engine bearings and gearboxes, are prone to mechanical wear, the effects of which are investigated in the domain of tribology. The presence of rotation and vibration may result in the continuous rubbing of the engine seals against each other during engine operation, leading to the removal of the base material and an increase in gaps. The effect of abrasive wear may be amplified through the cyclic operation (acceleration and deceleration) of the engine and may promote leakages and mechanical wear.

4.2. Thermal Distress

The combustor and turbine are the most common parts subjected to extremely high temperatures either directly by exposition to hot flow path gases or indirectly by their proximity to the engine's hot section. Stationary mechanical parts and rotating engine parts are common thermal distress-prone areas. These parts may include combustion chamber liners, turbine vanes, structural turbine cases, frames, turbine disks, turbine blades and rotating seals. Components, such as instrumentation devices for engine condition monitoring and fuel nozzles, are also affected by thermal distress.

Another identified mechanism is hot corrosion, which causes material loss of the affected component over time due to the chemical reaction between the base material and substances carried in the hot gas. This type of corrosion is also known as sulphidation, and the substance can originate from the fuel or from sources external to the engine, such as sulphates or salts. This type of corrosion, which is induced by a combination of sodium chloride from the inlet air and sulphur from the fuel, may have a detrimental effect on the integrity of hot section engine parts, such as high alloy HPT blades and vanes. High-temperature oxidation is another known distress mechanism, and it is induced by a chemical reaction between the base material and free oxygen from the hot gaseous environment. Removal of the material from the component is inevitable in this reaction. Furthermore, burn-off with significant material detachment can occur in the combustion and turbine sections due to excessive temperatures (Sallee, 1978). Immediate failure of the HPT blades will occur if spalling and removal of the thermal barrier coating (TBC) take place due to the exposition of the base material to high, beyond-melting-point temperatures.

With thermal distress, a common mechanism of deterioration is hot corrosion, which is material loss. Hot corrosion is caused by the chemical reaction between the base material and substances in the hot gas. Another known distress mechanism is high-temperature oxidation, which is caused by the chemical reaction between free oxygen from the hot gaseous environment and the base material. Similar to hot corrosion, it causes loss of materials in the engine parts (Ranier Kurz & Brun, 2007).

4.3. Abrasion, Corrosion and Erosion

Corrosion distress mainly occurs due to the chemical reaction between the base materials of parts and the environment. This phenomenon generally occurs due to the electrochemical oxidation of the exposed metal part reacting with oxygen from the surrounding air and/or moisture in the air. The cold section engine parts, such as the steel alloy LPC blades and vanes, are affected by this type of corrosion; their integrity is compromised.

Abrasion occurs as a result of material removal due to the rubbing of a moving blade tip against its static lining surface. It also occurs due to the grazing of a rotating inter-stage seal against its stationary counterpart. Flight loads and gyroscopic effects contribute to these types of graze and may cause engine shafts and cases to deflect from their designated location. Thus, the blade tip and seal clearances are decreased or increased. In addition, a high-temperature environment may result in the amplification of material contraction, which may induce abrasion in engine turbines.

Regarding part erosion, this degradation mechanism is a result of hard particles impinging a surface, thereby causing material reduction and diminishment of the part's original thickness. It is a common degradation in airfoils that are in direct contact with the path flow of air. An abrasive effect only becomes notable when the ingested particle is larger than 10 µm in diameter (Sallee, 1978). Common examples of these particles are dust, sand and other floating particles. Other stationary or rotating parts can be affected by erosion when exposed to air flow that carries abrasive particles. Cooling air passages and cavities within the engine in which circulation of air is constant may cause severe erosion.

Abrasion is defined as material removal due to rubbing. An example is the rubbing of a moving blade tip against its stationary lining surface or the rubbing of a rotating inter-stage seal against its stationary counterpart. Flight loads and gyroscopic effects cause rubbing, which in turn causes the engine shafts and cases to deflect from their initial positions. Therefore, the blade tip seal clearance is increased or decreased, which indicates a defect.

Corrosion distress generally occurs in parts by chemical reaction of the base material with the environment. Most corrosion cases occur when exposed metal parts react with the surrounding air or air with moisture. Cold section engine parts, such as alloy LPC blades, can lose their structural integrity with this type of corrosion.

Erosion is generally caused by hard particles impinging a surface, thereby rubbing away or replacing materials and diminishing the parts' initial thickness. Dust and sand are the most common causes of erosion. Erosion is common in airfoils because they are in direct contact with the air flow path. Common hard particles include dust, sand and other floating particles. Abrasive particles can also erode the stationary parts within the air flow path (Ranier Kurz & Brun, 2007).

4.4. Fatigue and Creep

4.4.1. Fatigue

Fatigue is caused by repeated alternating stress on a component in which cracks develop and grow with increased load cycles. The number of cycles causing failure ranges from less than 10 to 10^8 cycles and beyond. Fatigue cracks initiate in the region exposed to high stress concentration. Generally, this area is located at the root of any notch.

The growth rate of fatigue crack behaviour is shown in Figure 2. The fatigue crack growth rate can be divided into Regions I, II and III. Most of the component life is spent in Region I. Growth rates become extremely sensitive to mean stress and significant effects of the microstructure and environment. In this region, short cracks may grow faster that long cracks. Region II shows a linear behaviour of the rate, and Paris Law is obeyed. In this region, elastic modulus is the only material variable that influences crack growth rate. The effects of microstructure, mean stress and strength level are minimised. In Region III, the crack growth rate accelerates and approaches failure condition. Failure is dependent on the fracture toughness of the material. However, a substantial effect of mean stress is observed in this region.

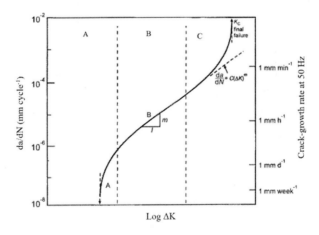

Figure 40. Fatigue crack growth rate (Ritchie, 2003).

The ranking of materials is reversed according to their stiffness, with aluminium being the worst and steel being the best in terms of growth rate. Growth rate varies with materials due to the substantial effect of the microstructure, environment and mean stress. Materials with high toughness allow large crack lengths to be present prior to catastrophic failure. This condition is often beneficial in increasing the probability of crack detection prior to catastrophic failure.

Fatigue is a major contributor to the failure of aircraft components. Statistics show that 60% of the total failure in aircraft components is caused by fatigue. The factors that influence the fatigue life of components are complex stress cycles, component engineering design, manufacturing and inspection of the components, service conditions and environment and material used in component constructions (Bhaumik, Sujata, & Venkataswamy, 2008).

4.4.2. Creep

Creep is a condition where the mechanical strength of metals decreases with increasing temperature. At high temperatures, atoms and dislocations become highly mobile, and this process is time dependent. Strength may include stress (σ), shear (ζ) and strain (ε). At room temperature, stress is a function of strain only (Eq. 1), whereas at elevated temperatures, stress is a function of strain, temperature and time (Eq. 2).

Room temperature:

$$\sigma = f(\varepsilon) \tag{1}$$

Elevated temperature:

$$\sigma = f(\varepsilon, t, T) \tag{2}$$

Surface oxidation may occur at high temperatures, slowly converting the structural material into brittle oxide. Different materials creep at different temperatures depending on the melting point of the materials (T_m).

For example, metallic metals creep when

$$T > 0.3 - 0.4\, T_m. \tag{3}$$

Ceramic creeps when

$T > 0.4 - 0.5\, T_m.$ (4)

Nickel base superalloys do not creep until

$T > 0.6 - 0.7\, T_m.$ (5)

5. DETERIORATION IN AIRCRAFT ENGINE COMPONENTS

The deterioration of components in a turbofan engine is a combined effect of all the mechanisms outlined in the previous section, namely, thermal distress, mechanical wear, corrosion, abrasion and erosion. The following subsection outlines the causes of component degradation and provides a quantitative assessment of the amount of degradation experienced.

5.1. Deterioration in the Compressor

LPC blades are one of the key components of gas turbine engines, and their main function is to convert energy. LPC deteriorations are caused by increases in tip clearances, increased airfoil surface roughness and blunting of the fan blade's leading edges. The fan blade's tip clearance increases with engine usage due to blade tip and casing wear. Most casings are equipped with wear strips to allow break in wear and prevent damage to the blades. However, additional blade growth occurs due to flight loads and transient operations, thus producing a gap larger than what is required for steady-state operations.

In addition, the wear strips experience erosion, thereby further increasing the clearances. Engine testing has established that tip clearance increases cause a reduction in compressor flow capacity and compressor efficiency (Naeem, 1996). This condition reduces the compressor's surge margin. Surface roughness, which is caused by the impact of erosive particles, also adversely affects compressor performance. A study showed that a 10% increase in airfoil roughness contributes to a 1% loss in compressor efficiency.

Figure 41. Compressor blade foul due to a mixture of oil and salt (Meher-Homji, Bromley, & Stalder, 2013).

Furthermore, roughness builds up rapidly (within the first 1000 cycles) then remains relatively constant. Particulate matter entrained into the engine also causes blunting of the leading edges of the compressor blades; the resulting change in airfoil shape leads to a decrease in compressor efficiency (Little, 1994; MacDonald, 1993). In summary, for JT9D, compressor deterioration is dictated primarily by tip clearance increases, surface roughening and airfoil contour changes. The combination of these loss mechanisms results in a decrease in compressor flow capacity and efficiency (Little, 1994).

The deterioration of HPC is qualitatively like that of LPC. Performance loss is associated with changes in tip clearances, rub strip erosion, blade length loss and airfoil erosion. The combined effect of these four deteriorations exceeds that of the fan; the total flow capacity and efficiency losses are as high as 10% and 8%, respectively (Kellersmann, Reitz, & Friedrichs, 2014; MacDonald, 1993). Figure 3 shows an example of deterioration in the compressor blade due to oil deposition and mixture of oil and salt.

5.2. Deterioration in the Combustor Chamber

The combustion chamber is a critical part of the gas turbine engine because it is exposed to high-temperature rise and prone to thermal distress due to combustion instabilities. The maximum combustion temperature is achieved when an ideal mixture between hydrocarbon fuel molecules and air occurs (Rotaru, 2017). The reaction process between fuel and air combustible mixture inside the combustion chamber releases energy, which is transferred to the surrounding, thereby causing temperature rise and oscillations of heat flux in the liner solid domain (Matarazzo & Laget, 2011). The maximum temperature that can be achieved in the combustion chamber is approximately 2000 K (Schulz, 2002). The mechanical strength of combustion base materials and coating declines rapidly due to this high temperature, thus damaging the chamber wall. Common failures in combustion liner involve degradation of material properties, bulging of the chamber, crack development and release of damaged pieces downstream. Protection systems, such as TBC and air cooling passages, have been developed to protect the liner and maintain the surface temperature below acceptable levels (Matarazzo & Laget, 2011).

Deterioration in the combustor chamber may involve choking of the fuel nozzles, which changes the fuel spray pattern, and/or combustor casing distortion, which may alter the critical dimensions of the combustor. The study conducted by Szczepankowski and Szymczak (Szczepankowski & Szymczak, 2016) showed that aviation fuel can become a source of soot and coke formation in the combustion chamber because aviation fuel contains a high percentage of resins and aromatics (approximately 28%). During the combustion of air and fuel mixture, large hydrocarbon particles are cracked into those with a small number of carbon atoms. Incomplete combustion of the particles forms resins, coke and ash (carbon deposits), as shown in Figure 4. The accumulation of incomplete combustion products, especially on the injector surface, not only results in changes in its operation condition (such as changes in spray angle) but can also lead to permanent damage (Szczepankowski & Szymczak, 2016).

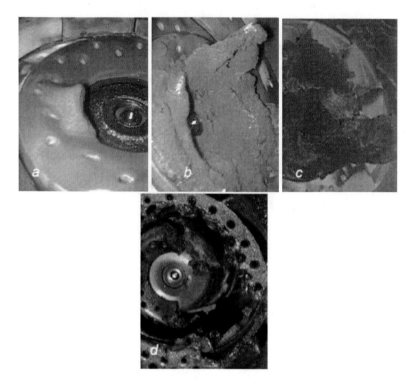

Figure 42. Build-up of incomplete combustion products (coke) in the air turbulator (Szczepankowski & Szymczak, 2016).

Serious carbon deposition in the combustion chamber results in performance degradation of the combustion chamber. Deposition can cause an uneven temperature distribution field at the exit of the combustion chamber. This increases the risk of overheating in certain local areas, thereby deteriorating the combustion chamber wall, turbine blade and the wall of thermal barrier coatings. Therefore, carbon deposition in the combustion chamber can seriously affect the normal operation of the engine and even the safety of the engine. However, no effective online monitoring method is available.

From a performance perspective, the two parameters of interest are the resulting overall pressure loss and combustor efficiency (Little, 1994). However, these parameters remain relatively invariant despite engine usage (Sallee, 1978). Therefore, combustor deterioration does not directly affect the overall performance loss. Nevertheless, these same deteriorations indirectly affect turbine performance. If the fuel nozzle spray pattern changes or the combustor casing is distorted due to flight loads, then the temperature profile, as seen by the turbine, will change (MacDonald, 1993).

5.3. Deterioration in the Turbine Blade

The turbine blade is the key component in aero-engine gas turbines. Turbines work under high-speed, high-pressure working environments, causing it to suffer from a more complex loading compared with other components in the aero-engine (Zhang, Lin, Zhang, & Liu, 2012). Moreover, the turbine blade experiences the highest combustor exit temperatures of approximately 2000 K. The most widely used turbine is made of nickel- or cobalt-based alloys that can withstand a maximum temperature of 1200 K [3]. Therefore, operating at a temperature that exceeds the maximum temperature degrades the turbine blade.

Mazur et al. (Mazur, Luna-Ramírez, Juárez-Islas, & Campos-Amezcua, 2005) listed the severe conditions of turbine blade as follows:

1. Operation environment, such as high temperature, fuel and air contamination and solid particles
2. High mechanical stress due to centrifugal force, vibratory and flexural stress
3. High thermal stress due to thermal gradients

The turbine blade experiences damage due to the severe conditions listed above, and these damages reduce its lifetime. Examples of turbine blade damages were listed in (Mazur et al., 2005) and shown below.

1. Creep
2. Thermal fatigue
3. Thermomechanical fatigue
4. Corrosion
5. Erosion

6. Oxidation
7. Foreign object damage

In addition to these severe conditions, the majority of turbine blade deterioration is also caused by (i) an increase in blade clearance, (ii) vane twisting and bowing and (iii) surface roughness of the blades. Increased clearances between the rotor blades and casing are caused predominately by centrifugal and thermal loads imposed during engine transients and by distortions of the engine casing as a result of changing flight loads (Kellersmann et al., 2014). Surface roughness in the turbine blade is caused by erosion. Erosion is a process in which the turbine blade is affected by solid particles on its surface. The eroding particles' characteristics are affected by temperature and impact conditions. Figure 43 shows an erosion in a turbine vane. This process increases heat transfer on the turbine blades. Compared with other loss mechanisms, surface roughness has a very minimal effect on turbine performance.

Figure 43. Erosion in the turbine vane (Hamed, Tabakoff, Rivir, Das, & Arora, 2004)

Vane distortion results from aerodynamic loads (arising from gas pressure bending moments and centrifugal untwist and loads) and stresses due to thermal gradients. As the vanes distort, coolant air is allowed into the main gas stream, resulting in a reduction in the turbine's efficiency (MacDonald, 1993). The magnitude of the resulting leakage can be as high as 2% of the flow, leading to an efficiency drop of up to 1.5% (Kellersmann et al., 2014). Bowing of the HPT's vanes causes the flow area and flow capacity to increase (Little, 1994). Most of the performance loss due to tip clearance increase occurs in the first 500 flight cycles. Subsequently, the clearances remain relatively invariant with the majority of flow and efficiency deteriorations occurring as a result of blade twist and bowing (MacDonald, 1993).

The deterioration in the LPT blade can be seen through clearance changes, twisting and bowing of turbine vanes and soldering of the vane's inner diameter. Vane soldering is caused by the misalignment of the vane's inner platform. This misalignment results in steps in the inner flow path surface, thereby causing aerodynamic losses and a loss in turbine efficiency. The airfoil's surface roughness increases with engine usage, but its effects are minimal compared with those of the other loss mechanisms (MacDonald, 1993). Twisting and tilting of the inner platform relative to the fastened-in-place outer platform result in the leakage of coolant air into the main stream gas. The result is a reduction in the LPT's efficiency (Kellersmann et al., 2014; Little, 1994). LPT may suffer from vane bowing, which causes the gas-path flow area to change, resulting in a decrease in flow capacity (Little, 1994). Other deterioration in turbine blade also is shown in Figure 44.

6. Effect of Deterioration on Engine Performance

Degraded engines, those that have operated for a substantial amount of time, show higher fuel consumption than the initial fuel consumption values at the test bed right after production. This condition means that the engine SFC increases over time due to the deterioration of the engine component efficiencies, mainly the compressor, combustor and turbines (Venediger, 2013). A commonly used parameter to describe the current engine condition is EGTM. This value is calculated by subtracting the maximum allowable EGT provided by the engine manufacturer from the actual EGT measured during engine operation.

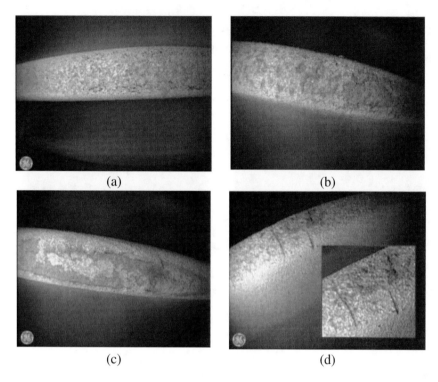

Figure 44. Turbine blade forms of damage: (a) micro-crack coating, (b) partial burn-out of the coating, (c) total burn-out of the coating until alloy is revealed, (d) partial burn-out of the coating and numerous macro-cracks (Błachnio, Kułaszka, & Zasada, 2015)

The maximum allowable EGT (EGT redline) represents the limit established by the engine manufacturer during certification tests and marks the maximum acceptable temperature at which the engine can operate without suffering from rapid deterioration. Peak EGT values are usually reached at or shortly after take-off and thus depend on the outside air temperature. The effect on the EGT margin on clean and deteriorated engines is shown in Figure 45. In summary, the deteriorated engine has a lower EGT margin than the clean engine and thus operates with decreased performance margins. Compressor fouling, seal leakage, increased tip clearances and airfoil erosion are examples of deterioration in the EGT margin. Deterioration in the EGT margin causes increases in engine thrust, decreases in engine usage, decreases in average flight distance and difficulty of the working environment (Yildirim & Kurt, 2018).

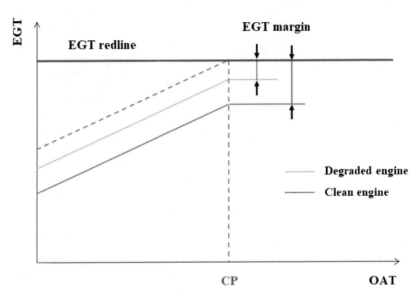

Figure 45. Comparison of the EGTM of clean and degraded engines (Venediger, 2013).

The effect of deterioration on overall aircraft engine performance can also be seen in terms of increment in turbine entry temperature (TET). Increments in TET are required to maintain the same level of thrust as that in the clean condition. The effect of compressor fouling can be clearly seen during take-off and climbing where the thrust setting is the highest, particularly in a short-haul mission. For a long-haul mission, the influence of degradation on TET increments is shown only during take-off. However, in both missions, deterioration in terms of compressor fouling increases the amount of fuel burnt (Igie, Goiricelaya, Nalianda, & Minervino, 2016).

Deterioration in engine components also exerts a significant impact on pressure changes across compressor stages. Research has shown that the pressure ratio of HPC decreases by approximately 10%, resulting in a 7% reduction in flow; this impact is associated with erosion in airfoils and depositions on the surface (Kellersmann et al., 2014). Sand erosion also causes losses in engine efficiency and engine performance. Solid particles suspended through an axial flow compressor cause changes in airfoil geometry, quality of the surface and changes in flow field. Non-uniform radial and circumferential distributions at the exit level and on the blade surface have been found to be the reason for changes in the flow field of compressor air (Tabakoff, 1987). Ghenaiet et al. (Ghenaiet, Tan, & Elder, 2004) also discovered losses in aerodynamic performance due to sand erosion. The consequences of these phenomena lead to a significant reduction in engine efficiency and engine performance. Depletion in engine efficiency is observed when a deteriorated engine flies at an elevated altitude (Koh, Mazlan, Rajendran, & Ismail, 2018). The deteriorated engine is required to burn more fuel to provide the required thrust for sustaining the aircraft speed because at a high altitude, the effect of density is dominant.

ACKNOWLEDGMENTS

The authors would like to thank USM Research University Grant (Grant No: 1001/PAERO/8014019), USM Bridging Grants (Grant No: 304/PAERO/6316111) and (Grant No: 304/PAERO/6316194) for the support to this study.

REFERENCES

Bhaumik, S., Sujata, M., & Venkataswamy, M. (2008). Fatigue failure of aircraft components. *Engineering Failure Analysis, 15*(6), 675-694.

Błachnio, J., Kułaszka, A., & Zasada, D. (2015). Degradation of the gas turbine blade coating and its influence on the microstructure state of the superalloy. *Journal of KONES, 22*.

Bräunling, W. J. (2015). *Aircraft engines: foundations, aero-thermodynamics, ideal and real circuit processes, thermal turboma-chinery, components, emissions and systems*: Springer-Verlag.

Diakunchak, I. S. (1992). Performance deterioration in industrial gas turbines. *Journal of engineering for gas turbines and power, 114*(2), 161-168.

Döring, F., Staudacher, S., Koch, C., & Weißschuh, M. (2017). Modeling Particle Deposition Effects in Aircraft Engine Compressors. *Journal of Turbomachinery, 139*(5), 051003.

Ghenaiet, A., Tan, S., & Elder, R. (2004). Experimental investigation of axial fan erosion and performance degradation. *Proceedings of the Institution of Mechanical Engineers, Part A: Journal of Power and Energy, 218*(6), 437-450.

Grieb, H. S. (2004). *Design of Turboprop Engines*: Basel-Boston-Berlin.

Hamed, A. A., Tabakoff, W., Rivir, R. B., Das, K., & Arora, P. (2004). Turbine Blade Surface Deterioration by Erosion. *Journal of Turbomachinery, 127*(3), 445-452. doi: 10.1115/1.1860376.

Hyde, C. J., Sun, W., & Hyde, T. (2011). An investigation of the failure mechanisms in high temperature materials subjected to isothermal and anisothermal fatigue and creep conditions. *Procedia Engineering, 10*, 1157-1162.

Igie, U., Goiricelaya, M., Nalianda, D., & Minervino, O. (2016). Aero engine compressor fouling effects for short-and long-haul missions. *Proceedings of the Institution of Mechanical Engineers, Part G: Journal of Aerospace Engineering, 230*(7), 1312-1324.

Kellersmann, A., Reitz, G., & Friedrichs, J. (2014). Numerical investigation of circumferential coupled deterioration effects of a jet engine compressor front stage compared to BLISK geometry. *Procedia CIRP, 22*, 249-255.

Koh, W. C., Mazlan, N. M., Rajendran, P., & Ismail, M. A. (2018). A Computational Study to Investigate the Effect of Altitude on Deteriorated Engine Performance. Paper presented at the *IOP Conference Series: Materials Science and Engineering*.

Kurz, R., & Brun, K. (2007). Gas Turbine Tutorial-Maintenance And Operating Practices Effects On Degradation And Life. Paper presented at the *Proceedings of the 36th Turbomachinery Symposium*.

Kurz, R., & Brun, K. (2012). Fouling mechanisms in axial compressors. *Journal of engineering for gas turbines and power, 134*(3), 032401.
Li, Y. (2002). Performance-analysis-based gas turbine diagnostics: A review. *Proceedings of the Institution of Mechanical Engineers, Part A: Journal of Power and Energy, 216*(5), 363-377.
Little, P. (1994). *The effects of gas turbine engine degradation on life usage.* MS thesis, Cranfield University, UK.
MacDonald, S. (1993). *A dynamic simulation of the GE F404 engine for the purpose of engine health monitoring.* MS thesis, Cranfield University, UK.
Matarazzo, S., & Laget, H. (2011). Modelling of the heat transfer in a gas turbine liner combustor. *Chia Laguna, Italy.*
Mazur, Z., Luna-Ramírez, A., Juárez-Islas, J. A., & Campos-Amezcua, A. (2005). Failure analysis of a gas turbine blade made of Inconel 738LC alloy. *Engineering Failure Analysis, 12*(3), 474-486. doi: https://doi.org/10.1016/j.engfailanal.2004.10.002.
Meher-Homji, C., Bromley, A., & Stalder, J.-P. (2013). Gas Turbine Performance Deterioration and Compressor Washing. Paper presented at the *Middle East Turbomachinery Symposia. 2013 Proceedings.*
Naeem, M. (1996). *Fuel usage and its effects on operational performance for GE-F404/F-18.* M. Sc. thesis, Cranfield University, UK.
Richardson, J., Sallee, G., & Smakula, F. (1979). *Causes of high pressure compressor deterioration in service.*
Ritchie, R. O. (2003). 4.14 - Fatigue of Brittle Materials. In I. Milne, R. O. Ritchie & B. Karihaloo (Eds.), *Comprehensive Structural Integrity* (pp. 359-388). Oxford: Pergamon.
Rotaru, C. (2017). Analysis of turbojet combustion chamber performances based on flow field simplified mathematical model. Paper presented at the *AIP Conference Proceedings.*
Roth, B. A., Doel, D. L., & Cissell, J. J. (2005). Probabilistic matching of turbofan engine performance models to test data. Paper presented at the *ASME Turbo Expo 2005: Power for Land, Sea, and Air.*

Ruichao, L., Yingqing, G., Nguang, S. K., & Yifeng, C. (2018). Takagi-Sugeno fuzzy model identification for turbofan aero-engines with guaranteed stability. *Chinese Journal of Aeronautics*.

Sallee, G. (1978). *Performance deterioration based on existing (historical) data*; JT9D jet engine diagnostics program.

Schulz, A. (2002). Convective and radiative heat transfer in Combustors. *High Intensity Combustors–Steady Isobaric Combustion*, 255-260.

Szczepankowski, A., & Szymczak, J. (2016). Initiation of damage to the hot part of aircraft turbine engines. *Research Works of Air Force Institute of Technology, 38*(1), 61-74.

Tabakoff, W. (1987). Compressor Erosion and Performance Deterioration. *Journal of Fluids Engineering, 109*(3), 297-306. doi: 10.1115/1.3242664.

Tarabrin, A., Schurovsky, V., Bodrov, A., & Stalder, J.-P. (1998). Influence of axial compressor fouling on gas turbine unit perfomance based on different schemes and with different initial parameters. Paper presented at the *ASME 1998 International Gas Turbine and Aeroengine Congress and Exhibition*.

Venediger, B. (2013). *Civil aircraft trajectory analyses-impact of engine degradation on fuel burn and emissions*.

Walsh, P. P., & Fletcher, P. (2004). *Gas turbine performance*: John Wiley & Sons.

Yildirim, M. T., & Kurt, B. (2018). Aircraft Gas Turbine Engine Health Monitoring System by Real Flight Data. *International Journal of Aerospace Engineering, 2018*.

Zhang, J., Lin, J., Zhang, G., & Liu, H. (2012). High cycle fatigue life prediction and reliability analysis of aeroengine blades. [journal article]. Transactions of Tianjin University, *18*(6), 456-464. doi: 10.1007/ s12209-012-1785-7.

Ziemianski, J. A., & Mehalic, C. M. (1980). *Investigation of performance deterioration of the CF6/JT9D, high-bypass ratio turbofan engines*.

In: Advances in Aerospace Science ... ISBN: 978-1-53615-689-8
Editors: Parvathy Rajendran et al. © 2019 Nova Science Publishers, Inc.

Chapter 6

IMPORTANT AERODYNAMIC PARAMETERS OF FLAPPING-WING UNMANNED AERIAL VEHICLES

N. A. Razak[*], Aizat Abas[2] and Zarina Itam[3]

[1]School of Aerospace Engineering, Universiti Sains Malaysia,
Pulau Pinang, Malaysia
[2]School of Mechanical Engineering, Universiti Sains Malaysia,
Pulau Pinang, Malaysia
[3]Department of Civil Engineering, Universiti Tenaga Nasional,
Putrajaya, Malaysia

ABSTRACT

Unmanned aerial vehicles (UAVs) with flapping wings have been the focus of many recent studies due to their numerous applications. This chapter emphasises the essential parameters for the development and evaluation of flapping-wing UAVs. These parameters should be considered to successfully design flapping-wing UAVs. This chapter provides a short description of several topics related to flapping-wing flight and covers the basics of unsteady aerodynamic force generation.

[*] Corresponding Author's Email: norizham@usm.my.

Subsequently, this chapter discusses important parameters of flapping-wing aerial vehicles, such as Reynolds number, flow unsteadiness, wing geometry and flapping kinematics. This chapter also presents the modelling of flapping wing aerodynamics using quasi-steady approaches.

Keywords: UAV, flapping wing, unsteady aerodynamics

1. INTRODUCTION

Unmanned aerial vehicles (UAVs) have various sizes, shapes, forms and types, such as fixed wing, helicopters, multirotor, flapping wing and lighter than air. These UAVs can perform multiple missions, such as surveillance, remote sensing and monitoring.

Amongst these types of UAVs, the flapping-wing UAV is unique, because it does not utilise rotating wings to generate thrust. Instead, thrust is generated by the flapping motion of the wings. Flapping-wing UAVs have different forms. The majority of these forms are based on biological organisms, such as birds, bats and insects. Flapping-wing UAVs inspired by birds are known as ornithopters, whereas those inspired by insects are called entomopters. Apart from their size, the differences between the two categories lie in the axis and frequency of flapping.

2. LIFT AND THRUST GENERATION IN FLAPPING-WING FLIGHT

Garrick (Garrick, 1937) showed that an airfoil section that is undergoing pure plunging sinusoidal motion in free stream velocity (U_∞) can generate thrust. Thrust generation is possible, because pure flapping motion induces a vertical velocity component that contributes to the presence of an effective angle of attack (α_{eff}) of a 2D wing. This phenomenon is identified as the Knoller (Knoller, 1909) and Betz (Betz, 1912) effect or the Katzmayr (Katzmayr, 1922) effect, as depicted in Figure 1. The α_{eff} of 3D flapping wings is a function of wing span and flapping period.

The unsteady aerodynamic forces encountered by a symmetric airfoil that is undergoing pitch and plunge oscillations are visualised in Figure 2. During downstroke, thrust is produced by tilting the wing downward, which results in a negative pitch angle. The vertical motion of the wing causes the relative incoming airflow (U_R) to approach from an angle below the wing. As the wing is pitched nose down, the tilt of the forces that result from the Knoller (Knoller, 1909) and Betz (Betz, 1912) effect is improved, and a considerable amount of thrust can be generated.

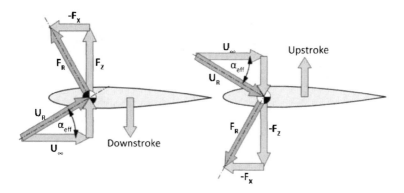

Figure 1. Knoller and Betz effect.

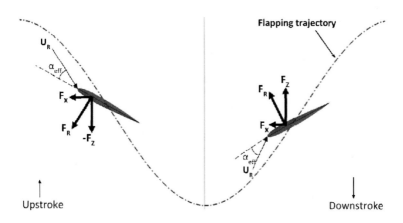

Figure 2. Aerodynamic forces and their corresponding α_{eff}.

For a symmetric airfoil, this phenomenon is true only if the pitch angle is lower than the relative incoming flow angle, thereby creating a positive α_{eff}. At a constant flapping frequency and incoming flow speed, an increasing pitch amplitude can be considered an increasing thrust, given that more tilt is achieved whilst reducing α_{eff}. This action is counterproductive, because it can reduce the total resultant force experienced by the wing, which decreases the amounts of thrust and lift. For the upstroke, the same principles (but in reverse) apply, thereby producing downforce instead of lift. Notably, thrust generation during upstroke is optional. Lift and drag may be generated during upstroke depending on the requirements. Some researchers have suggested that the wings of birds are aerodynamically inactive during upstroke. Birds use this technique to fly albeit only on the outboard wing region. Researchers believe that the inboard wing region of most large migratory birds produces lift and drag during both strokes under cruising flight condition, whereas the outboard section produces thrust during downstroke (Katzmayr, 1922)(Razak & Dimitriadis, 2014).

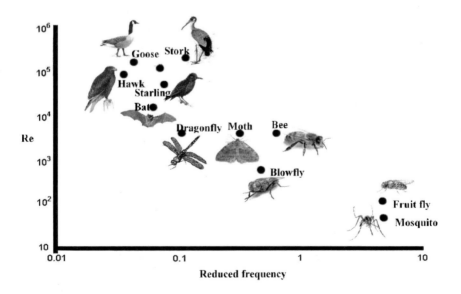

Figure 3. Re and reduced frequency of several biological flyers adapted from (Razak, 2012).

3. IMPORTANT FLAPPING WING PARAMETERS

Research related to flapping-wing flight typically requires an evaluation of several parameters. Key parameters include the Reynolds number (Re), flow unsteadiness, wing geometry and wing kinematics. Wing geometry can be divided into shape and size that correspond to wing planform, cross-sectional shape and morphing. The effects of these parameters are understood well in the field of steady-state aerodynamics. However, this knowledge cannot be readily applied to flapping-wing flight. Research on flapping flight has identified several key findings that contribute substantially to the understanding of flapping-wing flight.

3.1. Re

Figure 3 illustrates the Re range and flapping frequency of some biological organisms that utilise flapping wings to fly. The figure indicates a strong negative correlation amongst the parameters. Re, which is determined by wing chord size and forward flight speed, varies from tens to a few hundred thousand, whereas the maximum reduced frequency is approximately nine.

At a low Re, viscous force is dominant over inertial force. Rotating wings suffer from low efficiency at a low Re due to adverse viscous effects. Turbines and propellers become less efficient as the size and speed of a vehicle decrease. At a low Re, viscous drag increases due to a relatively thicker boundary layer, and flow separation causes the loss of lift and increased pressure drag (Chin & Lentink, 2016). To operate at a low Re, lift and thrust generation via flapping wings is the best option. A flapping wing can be designed such that the unsteady effects are successfully utilized to augment force production and efficiency at a low Re. Aerodynamic force generation can be enhanced by manipulating certain physical phenomena that only occur at a low Re, such as stable leading-edge vortex (LEV) and wake capture (Shyy, Lian, Tang, Viieru, & Liu, 2007). At a high Re, flow is irregular, unstable and consists of high-intensity vortex structures with a short temporal delay, thereby causing LEV to detach from the wing surface (Han, Chang, & Kim, 2014). Such observation indicates that a high Re can interrupt lift augmentation mechanisms, which is detrimental to a flapping-wing UAV.

3.2. Flow Unsteadiness

Reduced frequency represents the amount of unsteadiness experienced by a fluid when flowing around oscillating and flapping bodies. It describes the degree to which the transient phenomenon is important in a flow around an oscillating body at a specific frequency. The effects of this nondimensional parameter play a crucial role in determining the flow mechanism that produces lift and thrust. Birds and bats operate at relatively lower frequencies, but unsteady effects are also important in their flight mechanisms. In general, when the reduced frequency values exceed unity, the resulting flow is dominated by acceleration effects. In studies on flapping wings, the Strouhal number (St) can also be used as a measure of flow unsteadiness. This number describes the relative influences of flapping speed and amplitude on forward flight speed. It also characterises the wake vortex dynamics of a flapping wing in forward flight. Notably, a reduced frequency does not consider flapping amplitude (Razak, 2012). St can be evaluated using Equation 1.

$$S_T = \frac{fA}{U_\infty} \qquad (1)$$

Another parameter that is typically investigated by researchers in flapping flight research to characterise flow unsteadiness is advance ratio. This parameter incorporates the wingspan value into the equation, as shown in Equation 2. Advance ratio captures the 3D effect of flow, because a long wingspan experiences high wing tip velocities relative to the wing root (Dickson, 2004). The term $2\gamma Ts$ represents the total distance travelled by the wingtip from the start of the upstroke to the end of the downstroke. U_∞/f

implies the flapping wavelength, which denotes the forward distance travelled by the wing over a complete flapping cycle.

$$J = \frac{U_\infty}{2\gamma T s f} \qquad (2)$$

3.3. Wing Geometry

The effects of wing geometric parameters, such as airfoil thickness, leading-edge roundness, camber and aspect ratio, are well known in the field of steady-state aerodynamics. Some of this knowledge cannot be readily applied to flapping-wing flight, because the latter is not well understood. Wing camber improves lift generation and the maximal lift-to-drag ratio for wings operating at low and high Re values (Shyy et al., 2007). An efficient camber reduces the effect of drag. Okamoto and Elbina (OKAMOTO & EBINA, 2016) found that a high maximum lift coefficient can be obtained for a large camber circular arc airfoil, even at Re values lower than 3000. Furthermore, the pitching moment of the airfoil was determined to be zero within a wide range of angles of attack when the moment centre was set to an appropriate position. Wing models with sharp leading edges or wing camber have higher lift coefficients than wings without these features (Altshuler, Dudley, & McGuire, 2004).

The aerodynamic effect of wing thickness and leading-edge radius is strongly dependent on Re. A thick wing is advantageous at a high Re, because the range of acceptable attack angles becomes wider. Flow separation due to an adverse chordwise pressure gradient on the wing occurs considerably later than that on a thin wing (Shyy et al., 2007). By contrast, thin wings operate efficiently at a low Re. At a low Re, thin wings create more lift (Kunz, 2003) and less drag (Okamoto, Yasuda, & Azuma, 1996), which result in better lift and drag and increased performance. Some insects, such as dragonflies, have corrugated wing cross sections. Corrugated profiles and geometries with downward-facing leading edges produce slightly more lift. However, the effect is minimal and can be considered insignificant. The net vertical force of corrugated wing sections is approximately the same as that of the flat plate. The primary effect of corrugations is altering the pressure distribution on the windward side of the wing surface (Premachandran, S. and Giacobello, 2010). The influence of wing geometry on flapping wings appears unsettled, particularly at a low Re. Some findings contradict the discoveries from steady flow experiments.

3.4. Wing Kinematics

Wing kinematics deals with wing motion. In general, the motions of the wings of biological flyers are complex. The wings undergo translational and rotational motions in multiple axes. They bend, twist and morph simultaneously whilst undergoing flap, sweep fore and aft, twist, heave and pitch. Kinematic parameters vary from species to species. Large birds, such as albatrosses and migrating geese, operate at lower unsteadiness levels and utilise simpler wing kinematics compared with smaller birds, such as swifts and hummingbirds (Razak, 2012). Complexity is further increased by the fact that kinematics varies for different flight phases, such as landing, take off or manoeuvre. The current study did not attempt to describe complete kinematic possibility in flapping-wing flight.

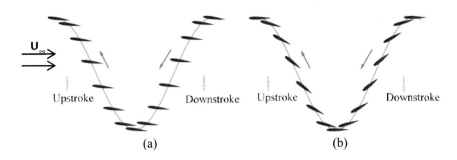

Figure 4. (a) Pure flapping and (b) flapping and pitching (lead).

The flapping cycle of a flapping wing is typically divided into two periods: downstroke and upstroke. In a case in which the stroke plane is perpendicular to the flight direction, downstroke is the period when the wing tip starts from its uppermost position and moves to its lowest position.

Upstroke begins from the wing tip's lowest position and ends when the wing tip has returned to its uppermost position.

One of the simplest flapping kinematics is known as the root flapping motion, which is shown in Figure 4(a). The motion involves the wing flapping up and down only in its flapping axis. Flapping and pitching are the motions by which the wing flaps and rotates simultaneously. Pitching indicates that the angle of the wing's cross section is maintained throughout the span wherein the pitch rotation axis is located. The cross section of a wing in flapping-and-pitching motion is shown in Figure 4(b). The phase angle between the two motions can vary such that the instantaneous pitch angle can lead or lag the instantaneous flap angle, as depicted in Figure 4(b). Many flapping-wing aerial vehicles, such as Delfly (de Clercq, de Croon, de Wagter, Ruijsink, & Remes, 2010) and SmartBird (developed by Festo), use flapping and twisting motions, because they generate higher lift and thrust with respect to the latter kinematics. Twist is produced by varying the pitch angle between the wing roots and the wing tips.

A twist is necessary, because the wing's cross section from the wing root to the wing tip differs in vertical tangential speed. Tangential speed V_Z along the wingspan can be obtained from the product of instantaneous angular flapping speed γ and the distance between spanwise positions to the rotational flapping axis r_s, as shown in Equation 3. Twisting optimises the relative angle of each wing cross section with respect to the incoming airflow.

$$V_Z = \dot{\gamma} r_s \qquad (3)$$

A close look on bird wings shows the existence of an elbow-like joint inside the wing. During flight, the elbow joint allows the wing to fold to minimise negative lift and drag generated during upstroke. The wing fold amount can vary. Many UAVs developed with bird-like flapping wings use a motor-driven four-bar mechanism that aims to achieve flapping motion. This mechanism can be modified to accommodate the elbow joint and wing twist (Razaami, Zorkipli, Lai, Abdullah, & Razak, 2017). Meanwhile, insect-inspired flapping-wing micro air vehicles can be driven by the piezoelectric effect to flap their wings (Peng, Cao, Liu, & Yu, 2017). Both driving mechanisms can be configured to handle tandem wing configuration.

Wing kinematics plays a major role in generating lift and thrust. The resulting unsteady aerodynamic forces around flapping wings can be optimised using a few special kinematics. Researchers have found four kinematic-related mechanisms used to amplify the amount of generated unsteady aerodynamic forces. These mechanisms are called high lift enhancement techniques. One of these enhancement techniques is LEV. LEV can be generated by operating at a high effective angle of attack, which can be achieved by flapping at a high frequency and/or increasing the pitch/twist angle. LEV is common for flight regimes in which Re is lower than 10,000. Experiments performed by van den Berg et al. (van den Berg & Ellington, 1997) and Ellington (Ellington, 1999) showed that an unsteady vortex bubble caused by flow separation from a sharp leading edge can explain the high lift characteristics of flapping flight. Insect flight (Ansari, Zbikowski, & Knowles, 2006) and hovering bird flight are considerably more dependent on LEV than other types of bird flight. The work conducted by Videler et al. (Videler, Stamhuis, & Povel, 2004) showed that LEV is also important for birds, such as swifts. Furthermore, a study by Muijres et al. (Muijres et al., 2008) reported that bats also use LEV to improve lift in slow flight regimes.

Another mechanism, called clap and fling (also known as the Weis–Fogh mechanism), was suggested by Weis–Fogh (Weis-Fogh, 1975) during an observation of the chalcid wasp (*Encarsia formosa*). A detailed examination of the clap-and-fling mechanism on insect wings can be found in (Lehmann & Pick, 2007). Notably, not all insects perform this kinematics. Clap occurs when the two wings are brought close together at the end of a stroke, thereby forcing the air between the gap to move out. The wings are then parted (fling) during the start of another stroke, thereby leading to the development of two vortices of equal strength but opposite signs on each wing. This mechanism has been shown to augment lift generation. Wakeling and Ellington (Wakeling & Ellington, 1997) also observed a clap-and-fling lift enhancement mechanism amongst dragonflies.

Fast wing rotation observed in insect wings also contributes to lift enhancement. The effect is similar to the Magnus effect. When a body moves through a viscous fluid whilst spinning, a boundary layer is generated around the body. This boundary layer induces fluid circulation around the body. Circulation is introduced when the velocity of the thin layer of fluid close to the body is slightly higher on a forward moving surface than on a backward moving surface, which experiences slightly lower velocity. This circulation amplifies static pressure on one side and lowers it on the other side. The variation in air pressure produces an aerodynamic force acting towards the low-pressure region. In the case of flapping wings during insect flight, the force acts normal to the wing chord. The timing of wing rotation during insect flight is important. If the wing flips before the end stroke, then the wing undergoes rapid pitch-up rotation in the correct translational direction, which enhances lift. This phenomenon is called advanced rotation. If wing rotation is delayed after the end stroke, then lift is reduced.

3.5. Wing Flexibility and Twist

The flexibility of flapping wings also affects unsteady aerodynamic force generation. A flexible wing adjusts its shape in response to external fluid forces, thereby affecting aerodynamic force production during flight (WOOTTON & KUKALOVÁ-PECK, 2007). The shapes of wing flexion are governed primarily by the inertial characteristics of the wing rather than the static pressure gradients arising from aeroelastic interactions. Aeroelastic feedback can play either a significant or negligible role depending on the elastic, aerodynamic and inertial loads acting on the wings. The wings may be regarded as purely inertial, flexible structures in a case where inertial forces are dominant.

Wing twist from a flexible wing structure is assumed to be essential for flapping flight, particularly for bird flight. However, wing twist exerts minimal effect on the forces generated by a flapping wing in insect-based flight. The unimportance of twist is attributed to the prominent role of unsteady aerodynamic mechanisms. Wing twist is the torsion of a wing parallel to the spanwise axis, thereby leading to a variation in the geometric angles of attack along the span. Twist allows the pitch angle to vary to compensate for the increasing tangential velocities along the span. Twist allows the wings to operate at a more or less constant effective angle of attack, which is close to the angle with the maximum lift-to-drag ratio (Razak, 2012).

4. Aerodynamic Modelling of Flapping Flight

Unsteady aerodynamic simulation can further promote the understanding of the unsteady flapping flight phenomenon. It can also be used to analyse the performance of flapping-wing aerial vehicles. Different simulation methods have been proposed, with varying degrees of complexity and fidelity. These simulation approaches can be used to evaluate unsteady aerodynamic efficiency generated by different flapping parameters. They can be classified into three wide categories: quasi-steady, unsteady (potential-based) and computation fluid dynamics. This article discusses only the quasi-steady approach, because the other two require a lengthy explanation.

Originally developed to estimate the performance of rotors, propellers and ducted fans, the actuator disc model developed by Rankine (Rankine, 1865) and Froude (Froude, 1889) is the simplest aerodynamic model for estimating the aerodynamic forces generated by flapping wings. This method is built on the assumption of a virtual actuator that maintains a pressure difference across itself by continuously pushing air, thereby causing a change in the momentum rate of the fluid, which leads to the possibility of estimating the resulting aerodynamic forces. The actuator represents the flapping wings for a flapping flight case. Ellington (Ellington, 1984) introduced a model for a partial actuator disc to match the flapping wing case. He modified the original theory by including non-uniform pressure distributions along with pulsing pressure to model flapping flight physics. Shkarayev and Silin (Shkarayev & Silin, 2010) described actuator disc theory as a relatively simple analytical treatment of the propulsive thrust and power of flapping flight. This method can provide only average quantities of flight performance based on a small number of inputs. The actuator disc model cannot predict instantaneous aerodynamic forces resulting from specific wing kinematics and geometries (Han et al., 2014).

Strip theory, also known as blade element momentum theory, is another quasi-steady method. This method works by dividing the wing into several spanwise panels of elements. The force is computed on each individual element using 2D steady-state aerodynamics. The elementary forces are then summed over the entire span to obtain the total forces generated by the wing. The calculation is repeated over the entire stroke cycle, which has been discretised into a series of time steps. The integration of the forces over a complete period yields the average force generated by the wings. Delaurier (DeLaurier, 1993) developed a modified strip theory to account for the wake effects. His approach allows for the camber and leading-edge suction effects to be included in the model. Post-stall behaviour can also be modelled. Moreover, spanwise static and dynamic twists can be included in the analysis. The theory is based on high-aspect-ratio wings that are largely undergoing chordwise flow, whereas spanwise flow is at the minimum. Delaurier's method yields better estimates of average forces, power requirements and propulsive efficiency. The method was used to predict the aerodynamic performance of pterosaur flight. Shyy et al. (Shyy, W. Kamakoti, R. Berg, M., and Ljungqvist, 2000) found that the predictions obtained using their own approach are comparable with those of Delaurier's results.

Unsteady wake effects are disregarded in quasi-steady theory. Flapping frequencies are assumed to be sufficiently low such that effects from shed wakes are considered negligible. Philip et al. (Phlips, East, & Pratt, 1981) applied this method to estimate the flight performance of birds during forward flight. Sane and Dickinson (Sane & Dickinson, 2001) argued that empirically modified quasi-steady state models can reasonably predict instantaneous aerodynamic forces due to the dynamic stall and rotational lift of flapping wings. However, Rayner (Raynerf, 1979) dismissed the idea that quasi-steady theory can predict the unsteady aerodynamics of flapping flight. Ellington (Ellington, 1984) demonstrated that quasi-steady analyses do not correctly predict the force magnitudes of the experimentally measured lift coefficient. The quasi-steady aerodynamic model can only be utilised in initial unsteady load estimation, because it disregards many unsteady effects.

CONCLUSION

This chapter establishes a general assessment of some key parameters of flapping flight. The understanding of flapping flight's vertical and horizontal force generation can be promoted by observing the 2D wing. The generation of vertical and forward horizontal forces can be easily described by breaking down the flapping stroke. Nevertheless, flapping wing kinematics is complex. It considerably affects the flow characteristics around flapping wings. Parameters, such as flow unsteadiness, wing geometry and Re, play major roles in the overall performance of flapping flight and should be studied. Furthermore, several lift enhancement mechanisms are essential for the flight capability of some biological organisms. These mechanisms can be applied to flapping-wing UAVs. Gradually, the best flapping-wing UAVs should mimic the mechanisms used by biological flyers to operate with high efficiency and manoeuvrability. The effects of these parameters can be evaluated by performing simulations using different aerodynamic models.

ACKNOWLEDGMENT

The author would like to acknowledge the support received through the Bridging Research Grant (304.PAERO.6316310) provided by Universiti Sains Malaysia.

REFERENCES

Altshuler, D. L., Dudley, R. & McGuire, J. A. (2004). Resolution of a paradox: Hummingbird flight at high elevation does not come without a cost. *Proceedings of the National Academy of Sciences*, *101*(51), 17731–17736. http://doi.org/10.1073/pnas.0405260101.

Ansari, S. A., Zbikowski, R. & Knowles, K. (2006). Aerodynamic modelling of insect-like flapping flight for micro air vehicles. *Progress in Aerospace Sciences*. http://doi.org/10.1016/j.paerosci.2006.07.001.

Betz, V. (1912). Ein beitrag zur erklarung des segel-fluges. *Zeitschrift Fur Flugtechnik Und Motor Luftschiffahrt* [A contribution to the explanation of the glider flight. *Journal of Aviation Technology and Motor Luftschiffahrt*], *3*(21), 71–73.

Chin, D. D. & Lentink, D. (2016). Flapping wing aerodynamics: from insects to vertebrates. *The Journal of Experimental Biology*, *219*, 920–923. http://doi.org/10.1242/jeb.042317.

de Clercq, K. M. E., de Croon, G. C. H. E., de Wagter, C., Ruijsink, R. & Remes, B. (2010). Design, Aerodynamics, and Vision-Based Control of the DelFly. *International Journal of Micro Air Vehicles*, *1*(2), 71–97. http://doi.org/10.1260/175682909789498288.

DeLaurier, J. D. (University of T. (1993). An aerodynamic model for flapping-wing flight. *Aeronautical Journal*, *97*(964), 125–130. http://doi.org/10.1017/S0001924000026002.

Dickson, W. B. (2004). The effect of advance ratio on the aerodynamics of revolving wings. *Journal of Experimental Biology*, *207*, The effect of advance ratio on the aerodynamics of. http://doi.org/10.1242/jeb.01266.

Ellington, C. P. (1984). The Aerodynamics of Hovering Insect Flight. V. A Vortex Theory. *Philosophical Transactions of the Royal Society B: Biological Sciences*, *305*(1122), 115–144. http://doi.org/10.1098/rstb.1984.0053.

Ellington, C. P. (1999). The novel aerodynamics of insect flight: applications to micro-air vehicles. *The Journal of Experimental Biology*, *202*(Pt 23), 3439–48.

Froude, R. E. (1889). On the part played in propulsion by differences of fluid pressure. *Transactions of the Institution of Naval Architects, 30,* 390–409.

Garrick, I. E. (1937). *Propulsion of a flapping and oscillating airfoil.*

Han, J. S., Chang, J. W. & Kim, S. T. (2014). Reynolds number dependency of an insect-based flapping wing. *Bioinspiration and Biomimetics, 9*(4). http://doi.org/10.1088/1748-3182/9/4/046012.

Katzmayr, R. (1922). *Effect of periodic changes of angle of attack on behavior of airfoils.*

Knoller, R. (1909). Die gesetze des luftwiderstandes. *Flug Und Motortechnik* [The laws of air resistance. *Flight and engine technology*], *3*(21).

Kunz, P. (2003). *Aerodynamics and design for ultra-low Reynolds number flight.* Stanford University.

Lehmann, F.-O., & Pick, S. (2007). The aerodynamic benefit of wing-wing interaction depends on stroke trajectory in flapping insect wings. *Journal of Experimental Biology, 210*(8), 1362–1377. http://doi.org/10.1242/jeb.02746.

Muijres, F. T., Johansson, L. C., Barfield, R., Wolf, M., Spedding, G. R. & Hedenstrom, A. (2008). Leading-Edge Vortex Improves Lift in Slow-Flying Bats. *Science, 319*(5867), 1250–1253. http://doi.org/10.1126/science.1153019.

Okamoto, M. & Ebina, K. (2016). Effectiveness of Large-Camber Circular Arc Airfoil at Very Low Reynolds Numbers. *Transactions of the Japan Society for Aeronautical And Space Sciences, 59*(5), 295–304. http://doi.org/10.2322/tjsass.59.295.

Okamoto, Yasuda. & Azuma. (1996). Aerodynamic characteristics of the wings and body of a dragonfly. *The Journal of Experimental Biology, 199*(Pt 2), 281–94.

Peng, Y., Cao, J., Liu, L. & Yu, H. (2017). A piezo-driven flapping wing mechanism for micro air vehicles. *Microsystem Technologies, 23*(4), 967–973. http://doi.org/10.1007/s00542-015-2762-6.

Phlips, P. J., East, R. A. & Pratt, N. H. (1981). An unsteady lifting line theory of flapping wings with application to the forward flight of birds. *Journal of Fluid Mechanics, 112*(1), 97. http://doi.org/10.1017/S0022112081000311.

Premachandran, S. & Giacobello, M. (2010). The effect of wing corrugations on the aerodynamic performance of low-Reynolds number flapping flight. In *17th Australasian Fluid Mechanics Conference*. Auckland.

Rankine, W. J. (1865). On the mechanical principles of the action of propellers. *Transactions of the Institution of Naval Architects, 6*, 13–29.

Raynerf, J. M. V. (1979). A New Approach to Animal Flight Mechanics. *J. Exp. Biol, 80*(1), 17–54.

Razaami, A. F., Zorkipli, M. K. H. M., Lai, H. C., Abdullah, M. Z. & Razak, N. A. (2017). Unsteady pressure distribution of a flapping wing undergoing root flapping motion with elbow joint at different reduced frequencies. *International Review of Aerospace Engineering, 10*(3). http://doi.org/10.15866/irease.v10i3.11530.

Razak, N. A. (2012). *Experimental investigation of the aerodynamics and aeroelasticity of flapping, plunging and pitching wings*. University of Liege.

Razak, N. A. & Dimitriadis, G. (2014). Experimental study of wings undergoing active root flapping and pitching. *Journal of Fluids and Structures, 49*. http://doi.org/10.1016/j.jfluidstructs.2014.06.009.

Sane, S. P. & Dickinson, M. H. (2001). The control of flight force by a flapping wing: lift and drag production. *Journal of Experimental Biology, 204*(15), 2607 LP-2626.

Shkarayev, S. & Silin, D. (2010). Applications of Actuator Disk Theory to Membrane Flapping Wings. *AIAA Journal, 48*(10), 2227–2234. http://doi.org/10.2514/1.J050139.

Shyy, W., Kamakoti, R., Berg, M. & Ljungqvist, D. (2000). A computational study for biological flapping wing flight. *Trans Aeronautical and Astronautical Society, 32*(4), 265–279.

Shyy, W., Lian, Y., Tang, J., Viieru, D. & Liu, H. (2007). *Aerodynamics of low reynolds number flyers. Aerodynamics Of Low Reynolds Number Flyers*. http://doi.org/10.1017/CBO9780511551154.

Van den Berg, C. & Ellington, C. P. (1997). The three–dimensional leading–edge vortex of a 'hovering' model hawkmoth. *Philosophical Transactions of the Royal Society of London. Series B: Biological Sciences*, *352*(1351), 329–340. http://doi.org/10.1098/rstb.1997.0024.

Videler, J. J., Stamhuis, E. J. & Povel, G. D. E. (2004). Leading-edge vortex lifts swifts. *Science*. http://doi.org/10.1126/science.1104682.

Wakeling, J. M. & Ellington, C. P. (1997). Dragonfly flight. II. Velocities, accelerations and kinematics of flapping flight. *Journal of Experimental Biology*, *200*(3), 557 LP-582.

Weis-Fogh, T. (1975). Unusual mechanisms for the generation of lift in flying animals. *Scientific American*, *233*(5), 81–7.

Wootton, R. J. & Kukalová-PECK, J. (2007). Flight adaptations in Palaeozoic Palaeoptera (Insecta). *Biological Reviews*, *75*(1), 129–167. http://doi.org/10.1111/j.1469-185X.1999.tb00043.x.

In: Advances in Aerospace Science ... ISBN: 978-1-53615-689-8
Editors: Parvathy Rajendran et al. © 2019 Nova Science Publishers, Inc.

Chapter 7

VISUAL LOCALISATION AND MAPPING USING UNMANNED AERIAL VEHICLES

Kai Yit Kok[1], Parvathy Rajendran[1,],*
Nurulasikin Mohd Suhadis[1] and Muhammad Fadly[2]
[1]School of Aerospace Engineering, Universiti Sains Malaysia,
Pulau Pinang, Malaysia
[2]Detrac Sdn Bhd, Hicom Glenmarie Industrial Park,
Selangor Darul Ehsan, Malaysia

ABSTRACT

This chapter presents important considerations that should be reviewed in unmanned aerial vehicle geospatial mapping missions: from preparation before the flight mission to the construction of a map model at the last stage. To obtain high-quality image data from missions, many preparations and analyses must be performed on hardware and software, especially for those who prefer low-cost flight missions with high accuracy. Safety should be the priority during flight missions to avoid injuries and casualties.

---

* Corresponding Author's Email: aeparvathy@usm.my.

Keywords: geospatial mapping, UAV, drone, mission, machine vision

1. INTRODUCTION

Unmanned aerial vehicles (UAVs) have elicited much interest among researchers due to their low cost, small platform and high manoeuvrability. They demonstrate a good usage potential in commercial and military fields, such as geospatial mapping (Achille et al., 2015).

A linear map and a digital terrain model (DTM) are obtained in traditional field surveying. DTM only provides the height information of a terrain surface. This conventional map is suitable for professional work because it removes features that are irrelevant to the purpose of the map and uses symbolic cartography to represent the objects on the map.

By contrast, UAVs can obtain data from a mission to produce various spatial products, such as image maps, a point cloud and a 3D textured map model. Products from UAV mapping, such as realistic 3D map models, are user-friendly because they illustrate all the features on the terrain with a realistic texture, thus enabling user interactions easily.

UAVs are also a good platform for efficient mapping missions because they can access dangerous or rough areas, especially mountainous ones that are inaccessible to field surveyors. Moreover, using UAVs may obtain more consistent and reliable image data without human errors compared with data obtained by field surveyors.

Although inconsistent images sometimes exist, these can be easily detected by observing overlap areas from other images adjacent to these images. With the assistance of an autopilot, the time required to complete the mapping task is much less than that in field surveying.

These advantages have made UAVs popular in geospatial mapping applications. However, several considerations, such as safety precaution, flight mission planning and the system for obtaining the desired mapping data, must be deliberated before starting a mapping mission using a UAV.

2. SAFETY PRECAUTIONS

Safety is the most important consideration in UAV geospatial mapping. UAVs should not be treated as toys but as cars or planes that can lead to injuries if controlled recklessly. Basic rules and regulations are applied when operating small civilian-purpose UAVs to ensure the safety of individuals and properties.

Given that UAV regulations and airworthiness are country-specific, the typical restrictions provided by the US Department of Defense (DOD) and the UK Civil Aviation Authority (CAA) are presented in this work (Haddon et al., 2002, 2004; Smith et al., 2014). Generally, no aircraft is permitted to fly across borders unless allowed by CAA. All aircraft that weigh less than 150 kg are required to fly (1) below 400 ft. from the surface (2) and within the sight of or less than 500 m away from the operator (3) at the maximum achievable speed of less than 70 kts.

Therefore, permission must be obtained from CAA prior to the flight of any small UAV for civilian purpose that is intended to be operated autonomously or piloted remotely beyond these limits. CAA also provides important guidance on the amount of kinetic energy in two conditions: free fall and total system failure flights.

Free fall is an unpremeditated but controllable descent because of partial system failure/s that require the aircraft to descend at 1.3 of its stalling speed in landing configuration at the maximum take-off weight. The other condition is when aircraft landing uncontrollable; here, the aircraft is required to fly at 1.4 of its maximum operating speed in landing configuration at the maximum take-off weight in the case of total system failure.

Under these conditions, without relying on any parachute deployment system, the aircraft should not exhibit kinetic energy exceeding 95 kJ. Ref. (Cunliffe et al., 2017) also prohibited the use of a metal propeller for either internal combustion engines or electric motors. Although small UAVs weigh less than 150 kg, operating or designing aircraft that weigh less than 7 kg is advantageous.

The most important element is exemption from expensive aircraft insurance by CAA, which helps reduce the operating cost, although insurance may be beneficial in the event of an accident. Consequently, during aircraft flight demonstration, a distance of at least 30 metres from the crowd line is required for an electric aircraft and 50 metres for a turbine-powered aircraft.

However, aircraft that weigh between 7 and 20 kg without fuel face the following additional constraints: (1) must obtain operating permission from air traffic control, (2) must have 50 ft of obstacle distance, (3) must be 30 metres away from the crowd line, (4) must possess adequate insurance coverage, (5) must have 150 metres of clearance from congested areas, (6) cannot perform aerobatics and (7) must have airworthiness certification issued by Ref. (Cunliffe et al., 2017). Currently, aircraft below 7 kg require no air traffic control permission unless operating in restricted airspace.

Despite this freedom, these aircraft are expected to obtain CAA permission before commencing operation in the future (Hodgkinson et al., 2018). Meanwhile, all UAVs developed for military, customs and police services are subjected to the regulations set by the Ministry of Defence regardless of their weight. Therefore, for civilian purpose, UAVs below 7 kg are highly encouraged.

3. MISSION PLANNING

Aside from safety, another main consideration in geospatial mapping using UAVs is deciding the UAV model for the mapping mission. The model can be of two main types, namely, fixed-wing and rotor UAVs. Each UAV model has a different performance in terms of flight speed, allowable payload weight, endurance, etc. Several researchers have studied the performance difference of different types of UAV models.

Ref. (Everaerts, 2009) analysed the performance of various airborne platforms, including their flexibility, positional accuracy, coverage and spatial resolution, which are relevant to remote sensing applications. Examples of fixed-wing UAV models are SwingletCAM made by SenseFly and Sirius I made by MAVinci. The HexaKopter model manufactured by MikroKopter is a multi-rotor UAV.

The most suitable UAV model is selected based on the mapping mission requirement. Basically, a multi-rotor UAV has less endurance than a fixed-wing UAV but can carry a heavier payload during flight. This feature enables its use in high-accuracy mapping missions because precise sensors and cameras with a large weight can be installed on it but within small areas that are less than a hundred hectares (Saadatseresht et al., 2015).

In comparison with rotor UAVs, fixed-wing UAVs have higher endurance and smaller weight, which make them preferable in large-area mapping missions over hundreds of hectares. In several cases, the mission requires an UAV to fly a pattern that is impossible for a fixed-wing UAV to follow. A rotor UAV is more appropriate for such a mission. Several mapping missions may only require the UAV to follow a smooth flight path; in this case, a fixed-wing UAV is more favourable.

The control mode of the flight for the mission (i.e., manual or autonomous mode) must also be decided. The control mode usually depends on the flight mission. The manual mode is preferred when the task focuses on monitoring or inspecting the area, whereas the autonomous mode is favourable for collecting static information on the target area.

Sometimes, the condition of the area may not be suitable for an UAV to fly autonomously for data gathering because of the presence of obstacles, such as large trees and power lines. Manual control mode is preferable in this case. Thus, investigating the target area is necessary before take-off.

However, no standard procedures exist for this task at the current stage, and examination through walking or driving around the area is the common approach. In addition, the flight mission should be simulated using available software with satellite imagery before the actual flight to obtain a better view of the mission.

Normally, UAV mappers use autonomous flight for mapping whilst the pilot is on standby to take control of the UAV during uncertain and deviating situations in real time. This scenario is crucial in terms of safety and the mission itself because cases exist where mission failure is not allowable despite long distances and extreme environment conditions in the target area.

UAV mappers also need to be alerted of any regulations or laws in the target area that prevent those who intend to fly drones in autonomous mode but are incapable of controlling UAVs manually. Having the skill before flying a drone in any area is still the safest way.

Although the autonomous control mode can minimise human errors and radio interference, which disturbs signals between the UAV and manual ground radio controller, it is not completely reliable because the UAV may behave abnormally at times due to bad weather, GPS interference or on-board technical errors during the flight.

In addition, complete trust must not be given to the software because no software in the world is bug-free, which is why software developers continuously apply software updates.

4. PAYLOAD SELECTION

Once mission planning, including the selection of UAV type and model, is completed, the crucial sensor in geospatial mapping (i.e., camera) should be determined based on the requirements of the image data to be obtained from the flight mission and the payload capacity limitation. Selecting a camera suitable for mapping may be difficult because many lightweight camera models are currently available in the market.

Several individuals choose a model with high performance in terms of photography or videography work, such as GoPro. The GoPro camera can undeniably capture a broad range of photos with its wide-angle lenses, but these lenses can distort images, thereby requiring extra work in eliminating distortion in the post-processing of images. Thus, cameras with wide-angle lenses are not recommended in geospatial mapping regardless of how well they capture videos or photos with high resolution.

When a rotorcraft-type UAV, such as a quadcopter, is used for mapping, the platform is highly unstable with obvious vibration by propellers and motors. A camera with a stable lens is recommended, and a camera with a lens retraction feature should be avoided, especially in mapping that requires high-accuracy measurements.

To overcome motion blur in images, an exposure time of 1/2,000 seconds or less is required from the camera (Rosnell et al., 2012) (Rosnell T., Honkavaara E. 2012). Furthermore, the bright areas in the images should not be saturated, and the shadowed areas should contain sufficient identifiable information. Hence, a camera with a high dynamic range is preferable.

High-grade cameras are generally heavier than common ones, and the payload restrictions of UAVs may lead to the use of low-grade cameras, which affect the image quality directly (Chiang et al., 2012; Rajendran et al., 2014). Hence, studies have been conducted over the years regarding the use of low-grade sensors whilst maintaining the image quality in mapping missions (Pfeifer et al., 2012; Wallace et al., 2012).

Cameras used for mapping can be usually programmed to allow pilots to control them manually or to be set at regular intervals based on the mission requirement. Several of them have internal GPS and can be programmed to capture pictures upon reaching certain coordinates. People can also obtain other types of images, such as thermal, from cameras with the help of specialised electronic sensors installed on UAVs.

Mobile phones with high-quality cameras that satisfy the mapping requirement have been developed. For example, the Nokia Lumia 1020 model has been used as a camera on LA300 UAV by a French company, Lehman Aviation. Moreover, MyungHyun Yun and his team investigated the use of Samsung Galaxy S and S2 smartphones mounted on a fixed-wing UAV to generate a digital elevation model (DEM) of an area located in South Korea (Leitão et al., 2015; Yun et al., 2012). Their results showed that the accuracy is slightly lower than that of expensive devices, but considering the cost and efficiency of smartphones, using them seems to be a reasonable option for those with a limited budget.

Cameras can be mounted on UAVs in several ways. Others fabricate a gimbal to hold the camera on the UAV because mapping only involves one or two camera angles, so the requirement for the gimbal is relatively simple. However, the performance of the camera with a simple home-made gimbal may be affected by turbulence and UAV vibration. To avoid this issue, a motorised gimbal is recommended to provide stabilisation for countering external disturbance and to allow the camera to capture clear images.

The integrity of sensor arrangement in UAV systems should be investigated before every flight mission to ensure that accurate data with negligible errors can be obtained from the system (Skaloud, 2007). UAV mappers can also test the performance and capability of UAVs in collecting high-accuracy geospatial data with simulation in a virtual environment to identify the optimum alignment of sensors in UAV systems.

The main purpose of mapping is to collect or capture image data. Thus, image quality is a crucial consideration in camera selection. Although the available flight planning software can perform all calculations for generating an optimum flight pattern that allows UAVs to obtain sufficient visual data, the image quality requirement for missions must be identified firstly.

Several mapping projects, such as aftermath disaster assessment, do not require high-resolution images, whereas others, such as archaeological mapping, require superior image quality. Image quality generally depends on camera specifications and flight altitude. We do not need to worry as much about flight altitude as camera performance because flight altitude can be adjusted to the optimum value by using a flight planning software.

Several camera specifications affect the image resolution, such as the focal length of the lenses, the digital sensor used (either a charge-coupled device (CCD) or a complementary metal-oxide-semiconductor (CMOS)) and the shutter speed. For example, CCD sensors can produce high-quality images with higher sensitivity and lower noise than CMOS sensors. Nevertheless, CMOS cameras have lower price and longer battery life than CCD cameras.

The measurement of image quality in geospatial mapping is performed with the ground sampling distance (GSD) unit. GSD is the distance between two consecutive pixel centres measured on the ground, or it can be referred to as the distance of one side of a pixel measured on the ground. For example, one-metre GSD means that the adjacent pixel location is one metre away from the ground. This means that a high GSD value entails low image quality and vice versa. Hence, flying at high and low altitudes using the same camera has high and low GSDs, respectively.

5. PAYLOAD INTEGRATION AND ATTACHMENT

After deciding on the camera model to be used in the mission, the next item UAV mappers need to examine is the orientation of the camera mounted on the UAV because the image data obtained from the camera differ according to the angle of the camera mounted on the UAV (Remondino et al., 2012).

Currently, nadir (overhead) and oblique are the two most commonly used aerial views in mapping. Nadir photograph is a traditional method in which the camera mounted on an UAV looks straight down, perpendicular to the horizontal level, capturing the top view of the subject. Although mapping using this method can produce a 3D map, the information on the objects in the map may be limited.

Oblique photograph is another means of gathering information on the landscape in 3D view by using the same camera but at an angle different from the straight-down mounted camera. Figure 1 illustrates the camera orientation in nadir and oblique views when installed on an UAV. The angle is not restricted (either high or low) as long as the angle remains the same throughout the flight; otherwise, image processing will become complicated and difficult. Oblique images provide a more comprehensive and detailed description of the target area than nadir images do (Rupnik et al., 2014).

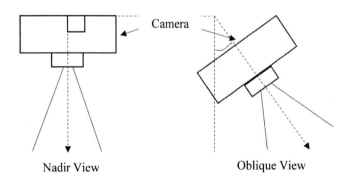

Figure 1. Illustration of nadir and oblique views.

3D imagery can be generated using a photogrammetry software, such as Agisoft PhotoScan, by using combined photos from the two methods. Many applications and simulations, including urban modelling simulation, accurate disaster simulation and post-disaster damage assessment, can be performed with this imagery.

6. FLIGHT ALTITUDE

When the hardware system has been configured, the appropriate flight altitude for the mapping mission must be determined before path planning for safety and legality concerns. Most UAVs available in the market are small. Therefore, the pilot must be prudent in controlling the altitude of the UAV within the line of sight. This law is enforced to prevent UAVs from flying at the same altitude level as manned aircraft to avoid mid-air collision. The UAV's pilot should be aware of the UAV altitude throughout the flight mission.

Furthermore, the flight mission requirement affects the flight altitude because several missions require UAVs to fly at low altitudes to capture high-resolution images, whereas others do not. Flying at low altitudes can increase the distortion in images, which is mainly caused by buildings or other objects on the ground.

Moreover, the time required to complete a mapping mission at low-altitude flight is much longer that that at high-altitude flight. Hence, the pros and cons of flying at different altitudes and the mission requirements and legality concern should be considered simultaneously to estimate the altitude in actual flight.

7. GEOREFERENCING

In geospatial mapping, the georeferencing process is always included in the preparation of the flight mission, especially for professional work. Georeferencing is often known as ground registration (Verhoeven et al., 2012) that explicitly defines the coordinate and rotation of the imagery corresponding to the Earth-related coordinate system.

In short, the selected points on a georeferenced map must be at the exact coordinates of actual geographical locations. Working with georeferenced maps is easy because these can be used on existing coordinates in software. However, georeferencing is time consuming, normally requiring an average of several minutes per locality (Guralnick et al., 2006).

Several areas have been georeferenced repeatedly, but the results remain inconsistent. To perform georeferencing using the common or traditional method, a certain number of ground control points (GCPs) must be identified in the beginning. GCPs are coordinate points of known locations on the Earth's surface.

The accuracy of the mapping mission must be determined without using GCPs, the accuracy of which can be 10–50 metres off and can be improved to within or less than 0.5–2 metres with GCPs (Safdarinezhad et al., 2015). With GCPs and GPS devices, images obtained from mapping can be georeferenced accurately.

The image data can then be used in mapping software, such as Google Maps, or can be transformed into a geographic control framework with various accessible geographic information system (GIS) tools, such as PCI Geomatica and ArcMap. Figure 2 shows an example of a georeferenced map.

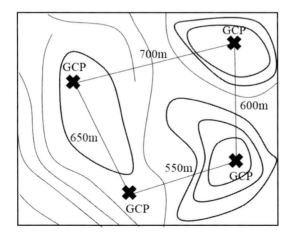

Figure 2. Georeferenced map (Kakaes et al., 2015).

Studies on improving the efficiency of georeferencing have been conducted to georeference a map without GCPs by proposing direct georeferencing (Nagai et al., 2009). This method can reduce and simplify mapping procedures (Lari et al., 2015). Normally, this is performed by deriving GPS and inertial measurement unit (IMU) data that can determine the position and orientation of the camera accurately. However, the positional accuracy of this method is usually lower than that of georeferencing with GCPs. Therefore, using GCPs when they are available remains the best option.

Advanced techniques for direct georeferencing include real-time kinematic (RTK) satellite navigation and the dual-frequency method, which can achieve tolerance within the centimetre range. These techniques outperform others, especially in mapping areas that lack available GCPs. However, the price of both is higher than those of other methods.

Nevertheless, in several cases, the priority of the mapping mission is not coordinate accuracy but the collection of general terrain information of the area for other applications. In these cases, the requirements for UAV software and hardware are simple and do not require spending time and money on methods and devices.

GCPs are not simply chosen randomly for georeferencing. The quality of GCPs can affect the accuracy of coordinates directly. Hence, several criteria should be considered when selecting good GCPs. Firstly, select features that can be identified accurately in the raw image and are as near as possible to the ground because several features, such as tall buildings, may be displayed at an angle and location from the top of the buildings and may probably deviate from the actual coordinates.

Secondly, repetitive features, such as lines on a highway, may be difficult to recognise in the image. Having GCPs in the overlap area between images is recommended because this can improve the accuracy of the images' coordinate system. In addition, having the GCPs scattered on the target area uniformly is ideal.

Generally, a large number of GCPs results in accurate data, and a minimum of five GCPs are normally necessary to perform georeferencing using a software. In several cases, UAV mappers may obtain GCPs from sources that have the same area with high accuracy, such as web maps (e.g., Google Earth) or laser scans.

For those who can afford to pay for or require accurate georeferencing, a global navigation satellite system (GNSS) device may be another option. This device was developed for use with GPS, GLONASS, Galileo or the Beidou system to provide precise coordinates. It has a precise crystal oscillator that can have positioning errors of one metre or less in real time by decreasing the errors induced by the receiver clock jitter. During post-processing, the positioning errors can be further reduced.

8. FLIGHT PATH PLANNING

In path planning, UAV mappers must determine the software to be used (either open-source or paid software). A common open-source software is Mission Planner. It can be used in fixed-wing and rotor UAVs, and the paid one is usually developed with hardware for a specific UAV model, such as the osFlexPilot developed by Airware Company for use on fixed-wing UAVs.

The functionality of most software is about the same. Open-source software is reliable and popular amongst people with a limited budget. Open-source software even has a community group whose members help each other regarding the software, and those with programming skills can contribute by debugging or upgrading the software.

In addition, research has been conducted on generating flight paths and determining waypoint locations automatically with only the initial and final coordinates of the flight by using an optimising algorithm (Guralnick et al., 2006; Kok et al., 2016; Kok et al., 2015), which can make flight planning software more user-friendly in the future.

The flight plan is prepared using a flight planning software through the following steps: the on-board flight controller is connected to the ground radio controller and linked to a computer through the Internet or a direct USB link. If Internet connection is available, the flight mission can be accomplished at the site; otherwise, UAV mappers must conduct flight planning and upload the flight plan to the flight controller before entering the target area.

During path planning, a flight path is designed in the software by defining the waypoints around the target area, the flight altitude, the settings of the control system and the camera settings (model and how to trigger photo shooting). Several considerations must be deliberated before determining the waypoint coordinates of the flight path, such as the presence of obstacles, terrain condition, flight lines that can provide an adequate overlap area between images and suitable take-off and landing locations if a fixed-wing UAV is used in the mission (Ghadage, 2014).

Then, based on the safety concerns and mission requirements, a certain number of waypoints are set on the map and linked to each other with straight lines to form a flight path. In a mapping mission, the flight path pattern is likely to have multiple parallel lines with waypoints on these lines within the target area, and the interval distance between the parallel lines must be smaller than the camera's coverage area to ensure that the images overlap each other. A short distance between parallel lines leads to a large overlap area between images and increases the flight path length.

When the flight plan and other settings have been set, the flight path and other useful information, such as the number of images required for the mission and GSD estimation, are generated by the software. Users can still modify several of the settings, including the amount of overlap area and the operational altitude, if they are not satisfied with the current parameters. Afterwards, the flight plan and other settings are transferred or uploaded to the flight controller through the connection between the computer and the flight controller.

During the flight, the window screen of the software on a computer or tablet shows in-flight data, such as GPS status, ground signal status, current coordinate and orientation of the UAV and camera view. The pilot can test the settings or readings received from the sensors on-board before the actual flight mission to ensure that the drone is working as expected.

The data are either saved to the flight controller and transferred to the computer after the flight or transferred directly to the computer at the ground station. If the flight is a long-distance mission, storing data to the flight controller may be the appropriate way to preserve the data safely throughout the flight.

To maximise image quality, several people suggest maintaining the UAV altitude at a constant level above the ground whilst ignoring the height of objects and the terrain condition in the area to obtain consistent image data. Others prefer to have two dissimilar flight patterns at different altitudes which overlap with each other over the same area. Although this method is time consuming and entails much work, an adequate amount of image data can be gathered at different altitudes, thus overcoming the elevation variation issue.

ACKNOWLEDGMENTS

The authors acknowledge the support provided by USM Fellowship and RU Grant Universiti Sains Malaysia 1001/PAERO/814276 to this work.

REFERENCES

Achille, Adami, Chiarini, Cremonesi, Fassi, Fregonese, & Taffurelli. (2015). UAV-based photogrammetry and integrated technologies for architectural applications—Methodological strategies for the after-quake survey of vertical structures in Mantua (Italy). *Sensors, 15*(7), 15520-15539.

Chiang, Tsai, & Chu. (2012). The development of an UAV borne direct georeferenced photogrammetric platform for ground control point free applications. *Sensors, 12*(7), 9161-9180.

Cunliffe, Anderson, DeBell, & Duffy. (2017). A UK Civil Aviation Authority (CAA)-approved operations manual for safe deployment of lightweight drones in research. *International journal of remote sensing, 38*(8-10), 2737-2744.

Everaerts. (2009). NEWPLATFORMS—Unconventional Platforms (Unmanned Aircraft Systems) for Remote Sensing. *Official Publication*(56).

Ghadage. (2014). *Novel Waypoint Generation Method for Increased Mapping Efficiency Using UAV.* Arizona State University.

Guralnick, Wieczorek, Beaman, Hijmans, & Group. (2006). BioGeomancer: automated georeferencing to map the world's biodiversity data. *PLoS biology, 4*(11), e381.

Haddon, & Whittaker. (2002). Aircraft airworthiness standards for civil UAVs. *UK Civil Aviation Authority (CAA), London, United Kingdom*.

Haddon, & Whittaker. (2004). UK-CAA policy for light UAV systems. *UK Civil Aviation Authority*, 79-86.

Hodgkinson, & Johnston. (2018). *Aviation Law and Drones: Unmanned Aircraft and the Future of Aviation*: Routledge.

Kakaes, Greenwood, Lippincott, Dosemagen, Meier, & Wich. (2015). Drones and aerial observation: New technologies for property rights, human rights, and global development. *New America, Washington, DC, USA, Tech. Rep.*

Kok, & Rajendran. (2016). Differential-evolution control parameter optimization for unmanned aerial vehicle path planning. *PloS one, 11*(3), e0150558.

Kok, Rajendran, Rainis, & Ibrahim. (2015). Investigation on selection schemes and population sizes for genetic algorithm in unmanned aerial vehicle path planning. Paper presented at the *2015 International Symposium on Technology Management and Emerging Technologies (ISTMET)*.

Lari, & El-Sheimy. (2015). System considerations and challendes in 3d mapping and modeling using low-Cost uav systems. *The International Archives of Photogrammetry, Remote Sensing and Spatial Information Sciences, 40*(3), 343.

Leitão, de Vitry, Scheidegger, & Rieckermann. (2015). Assessing the quality of Digital Elevation Models obtained from mini-Unmanned Aerial Vehicles for overland flow modelling in urban areas. *Hydrology & Earth System Sciences Discussions, 12*(6).

Nagai, Chen, Shibasaki, Kumagai, & Ahmed. (2009). UAV-borne 3-D mapping system by multisensor integration. *IEEE Transactions on Geoscience and Remote Sensing, 47*(3), 701-708.

Pfeifer, Glira, & Briese. (2012). Direct georeferencing with on board navigation components of light weight UAV platforms. *Int. Arch. Photogramm. Remote Sens. Spat. Inf. Sci, 39*, 487-492.

Rajendran, & Smith. (2014). The development of a small solar powered electric unmanned aerial vehicle systems. Paper presented at the *Applied Mechanics and Materials*.

Remondino, Del Pizzo, Kersten, & Troisi. (2012). Low-cost and open-source solutions for automated image orientation–A critical overview. Paper presented at the *Euro-Mediterranean Conference*.

Rosnell, & Honkavaara. (2012). Point cloud generation from aerial image data acquired by a quadrocopter type micro unmanned aerial vehicle and a digital still camera. *Sensors, 12*(1), 453-480.

Rupnik, Nex, & Remondino. (2014). Oblique multi-camera systems-orientation and dense matching issues. *The International Archives of Photogrammetry, Remote Sensing and Spatial Information Sciences, 40*(3), 107.

Saadatseresht, Hashempour, & Hasanlou. (2015). UAV photogrammetry: a practical solution for challenging mapping projects. *The International Archives of Photogrammetry, Remote Sensing and Spatial Information Sciences, 40*(1), 619.

Safdarinezhad, & Zoej. (2015). An optimized orbital parameters model for geometric correction of space images. *Advances in Space Research, 55*(5), 1328-1338.

Skaloud. (2007). Reliability of direct georeferencing-beyond the Achilles' heel of modern airborne mapping. Paper presented at the *Photogrammetric Week*.

Smith, & Rajendran. (2014). Review of the elementary aspect of small solar-powered electric unmanned aerial vehicles. *Australian Journal of Basic and Applied Sciences, 8*(15), 252-259.

Verhoeven, Doneus, Briese, & Vermeulen. (2012). Mapping by matching: a computer vision-based approach to fast and accurate georeferencing of archaeological aerial photographs. *Journal of Archaeological Science, 39*(7), 2060-2070.

Wallace, Lucieer, Watson, & Turner. (2012). Development of a UAV-LiDAR system with application to forest inventory. *Remote Sensing, 4*(6), 1519-1543.

Yun, Kim, Seo, Lee, & Choi. (2012). Application possibility of smartphone as payload for photogrammetric UAV system. *International Archives of the Photogrammetry, Remote Sensing and Spatial Information Sciences, 39*(B4), 349-352.

In: Advances in Aerospace Science ...   ISBN: 978-1-53615-689-8
Editors: Parvathy Rajendran et al.   © 2019 Nova Science Publishers, Inc.

Chapter 8

GEOSPATIAL MAPPING USING SATELLITES

Mohd Badrul Salleh, Nurulasikin Mohd Suhadis, Parvathy Rajendran and Hassan Ali*
School of Aerospace Engineering, Universiti Sains Malaysia,
Pulau Pinang, Malaysia

ABSTRACT

A spaceborne system and its elements for geospatial mapping missions are discussed in this chapter. Each element plays an important role in ensuring the accomplishment of such missions. Therefore, the need to define, select and design these elements appropriately is crucial. As space technology advances and matures, many design options have emerged. These options have bridged the gaps and reduced the limitations inherent in spaceborne remote sensing systems. Whether we are looking back into the past, realising the present or imagining the future, satellite platforms have already and will always contribute to geospatial studies to better understand our home, the planet Earth.

Keywords: remote sensing, space technology, orbit, data acquisition

---

* Corresponding Author's Email: nurulasikin@usm.my.

1. Introduction

Spaceborne geospatial mapping is common nowadays. Every day, dozens of satellites orbit the Earth, perform sophisticated Earth observation missions and send thousands of images to end users on Earth for various purposes. The use of satellite platforms to acquire images of the Earth's surface can be traced back to the 1960s during the Cold War when the United States had to perform reconnaissance missions on the Soviet Union's military strength without being detected (Baumann, 2009).

Since then, the term 'remote sensing' has been used, and the application of remote sensing satellites has rapidly extended from military and defence to numerous fields, such as topography, urban planning, geology, forestry and even disaster risk reduction and management, as shown in Figure 1.

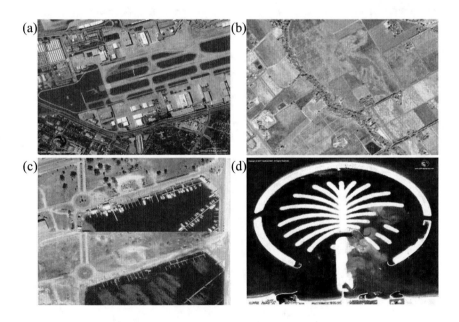

Figure 1. (a) WorldView-2 (0.5 m) image map of the Dallas Love Airport, (b) crop health variations captured by the QuickBird satellite, (c) satellite images before and after Hurricane Katrina in Louisiana and (d) satellite image used for the strategic urban planning of Palm Islands in Dubai (DigitalGlobe, 2013).

The need to use satellite platforms in geospatial mapping is firstly attributed to the capability of satellites to cover large surfaces for long periods without the concern of crossing airspace and political borders. Secondly, satellites can generate useful *in situ* data from regions of interest even with restrictions or insufficient resources.

As space system technology further advances, satellite platforms will continue to benefit geospatial mapping and the communities working in this field. To appreciate these opportunities, understanding how spaceborne systems and their elements for geospatial mapping missions are defined and designed before spatial images of Earth are successfully sent to end users is important.

This chapter is organised as follows. The operation concept of spaceborne geospatial mapping missions is discussed in Section 2. Section 3 describes the spacecraft as a platform for such missions. Section 4 explains the spacecraft's orbits. Section 5 discusses onboard payloads for geospatial mapping. Section 6 describes data acquisition techniques. Section 7 explains how acquired images are sent to ground stations. A case study of Landsat 8 is presented in Section 8. Lastly, Section 9 provides the conclusion of this chapter.

2. MISSION OPERATION CONCEPT

In general, the concept of geospatial mapping using satellites is similar to those of other methods, such as airborne systems using unmanned aerial vehicles (UAVs). However, unlike in the UAV platform, images are acquired by satellites from altitudes higher than 100 km, i.e., in space, at a specific orbit. Satellites used in geospatial mapping are known as remote sensing satellites and their primary mission is to perform Earth observations.

These satellites orbit around Earth and are equipped with payloads, i.e., onboard sensors that act as 'eyes' and 'ears', which sense the electromagnetic (EM) spectrum reflected off or emitted by objects of interest on Earth. As required by users, these satellites can be 'commanded' to acquire images along their flight track or at certain regions within a specific time. The acquired data are then processed by an onboard computer, which relays the data to users on the ground (Figure 2) for analysis, documentation and even commercialisation.

3. SPACECRAFT

UAVs are used in airborne remote sensing systems. By contrast, a spacecraft is a platform used to carry remote sensing payloads for Earth observation missions in space. This platform, also known as a satellite, consists of two main parts: the payload and the spacecraft bus (Figure 3).

The payload is the part of a spacecraft that performs the mission. For geospatial mapping missions, the payload can be any scientific instrument that can detect EM radiation from a target on Earth. Before selecting the payload, its requirements, such as power budget, types of EM signals to detect, target area, data resolution and the instrument's volume and mass, must be firstly defined. Defining a payload's requirements is crucial, because the payload influences the spacecraft bus and orbit design processes. Moreover, it determines the success of a mission. The payload system is discussed in Section 5.

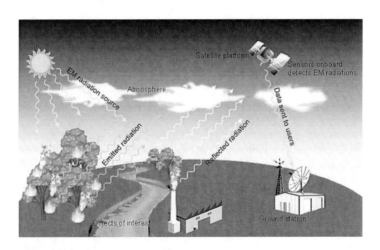

Figure 2. Mission operation concept of geospatial mapping using a satellite platform.

Geospatial Mapping Using Satellites

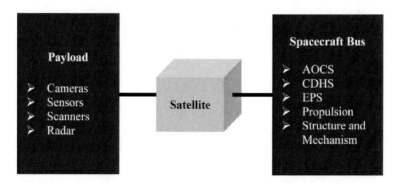

Figure 3. Spacecraft system that consists of payloads and several subsystem elements that comprise the spacecraft bus.

Meanwhile, the spacecraft bus is the subsystem that supports the payload to enable it to function. For example, the attitude and orbit control system (AOCS) determines and stabilises attitude and keeps the satellite in its orbit. The communication and data handling system (CDHS) is responsible for sending data to and receiving commands from a ground station. It also processes and stores data in an onboard computer. The electrical power system (EPS) has solar panels and a battery to generate and store electrical power, respectively. The EPS supplies power to the payload. Lastly, the structure and mechanism system provide support to mount and move the payload. All subsystem elements, as indicated in Figure 3, must be appropriately designed and integrated to provide optimal functions to the payload throughout the mission period of a satellite (Sellers et al., 2003).

4. ORBITS AND FORMATION FLYING

An orbit is the flight path of a satellite around Earth. To place a satellite into orbit, a launch vehicle (LV), which is either a rocket or a space shuttle (has been terminated), is used. At a certain altitude, the satellite is deployed from the LV. The orbit's altitude where the satellite will perform its mission depends on the mission and payload requirements. Several types of orbit are typically considered for remote sensing satellites.

4.1. Geostationary Earth Orbit (GEO)

GEO is a circular orbit with an altitude of approximately 35,789 km above the Earth's equator. It has the same orbital period as the Earth's rotational period, i.e., 24 hours. Therefore, any satellite in this orbit appears to be motionless at a fixed position in the sky and is always pointing towards the same region on Earth over a large coverage area. This orbit is suitable for weather and ocean monitoring satellites but is less preferable for high-resolution mapping missions, because it provides low spatial resolution due to its high altitude.

4.2. Low Earth Orbit (LEO)

LEO is an orbit with an altitude ranging from 160 km to 2000 km. This altitude range allows satellites to acquire high spatial resolution images at a frequent revisit period per day. Therefore, LEO is commonly considered for remote sensing satellites to accomplish Earth observation missions. Depending on the orbital inclination and altitude, LEO can be designed with certain properties that allow satellites to fulfil their missions. Examples of LEO are a Sun-synchronous orbit (SSO), a dawn/dusk orbit and a polar orbit.

4.2.1. SSO

SSO, also called a Heli synchronous orbit, is an orbit with an altitude ranging from 600 km to 800 km, an inclination of approximately 98° to the equator and an orbital period of 96–100 minutes, as shown in Figure 4. Its orbital properties place satellites in constant sunlight. Thus, this orbit is highly useful for Earth observation missions, in which the payloads onboard a satellite require sunlight to acquire image data. A high-resolution Earth observation satellite, called *GeoEye-1*, owned by DigitalGlobe, is an example of an Earth observation satellite that utilises this orbit.

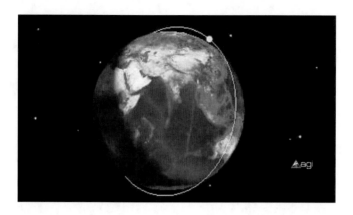

Figure 4. Example of an SSO (STK, 2019).

4.2.2. Dawn/Dusk Orbit

A dawn/dusk orbit is a special case of SSO. Its local mean solar time of passage for equatorial longitude is around sunset or sunrise. This orbit characteristic places satellites at the terminator of day and night, as illustrated in Figure 5. 'Riding' the terminator is useful for a satellite with payloads that require continuous and high-power supply, because the solar panels of the satellite can always be illuminated by the sun without being blocked by the Earth's shadow. In addition, this orbit is beneficial for a satellite that is always required to point its payloads towards the night side of the Earth to limit the sun's influence on measurements.

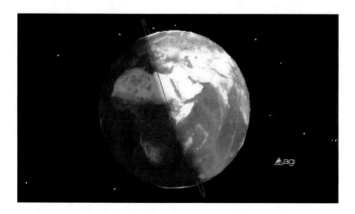

Figure 5. Satellite in SSO 'riding' the terminator of day and night (STK, 2019).

4.2.3. Polar Orbit

A polar orbit is a common type of orbit used in Earth observation missions. This orbit, which has an inclination of or nearly 90° to the equator, allows satellites to fly over the north and south poles. In addition, the satellites can hover above one of Earth's polar regions for a long period by using a highly elliptical polar orbit with its apogee above that region, as illustrated in Figure 6.

4.3. Satellite Formation Flying

Satellite formation flying is a concept that utilises a cluster of satellites, typically small satellites, that work together to accomplish missions. Flying multiple small satellites in formation offer flexibility to missions; they can cover a large surface area, acquire images from multiple angles at multiple times and reduce the cost, size and complexity of a large satellite (Sabol et al., 2001). This revolutionary concept can be implemented to compensate for problems that a single satellite in LEO has to face, i.e., short scanning interval over a small coverage area due to its high orbital velocity. On the basis of flight history, constellation and trailing formations are two types of satellite formation flying that have been considered for Earth observation satellites in LEO.

4.3.1. Satellite Constellation

A satellite constellation consists of a group of satellites, typically identical satellites, which orbit the Earth at the same or different orbital plane with coordinated ground coverage whilst performing the same mission, as shown in Figure 7. The flight path and position of these satellites are designed such that they are synchronised and overlap well in the coverage.

This type of formation flying increases the coverage area and allows a continuous coverage of the Earth's surface at any time, which a single satellite cannot accomplish. In addition, it helps reduce failure risk, because the remaining satellites can back up and continue to perform the mission if one of the satellites in the cluster fails. For example, the commercial remote sensing satellite, RapidEye, works in a group of five identical Earth observation satellites in the same orbital plane at an altitude of 630 km. The satellites can acquire image data from an area of over 4 million km^2 at a resolution of 5 m in five distinct bands of the EM spectrum. Figure 8 illustrates the satellite constellation formation of RapidEye, which acquires reliable information about agricultural crop monitoring and mapping.

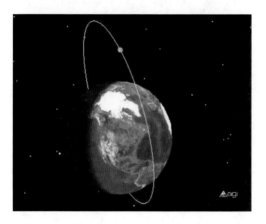

Figure 6. Polar orbit with an apogee above the North Pole that allows a satellite to hover for a long time above that region (STK, 2019).

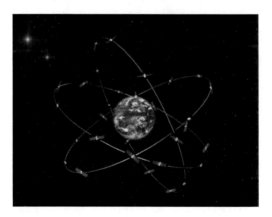

Figure 7. Constellation of identical satellites in the same and different orbital planes (Panissal et al., 2010).

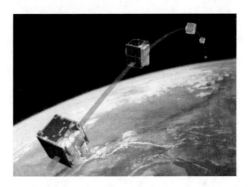

Figure 8. Five RapidEye satellites in constellation formation (Kramer, 2002).

4.3.2. Trailing Formation

A trailing formation is another type of formation flying that consists of at least two satellites orbiting on the same plane. The trailing satellite follows the leading satellite, and the two satellites are separated by a specific time interval. Therefore, objects of interest can be viewed from multiple angles at multiple times within a certain time interval. This formation is suitable for environmental observation and mapping, such as in monitoring the progress of wildfire and land use and in building a stereographic of terrain images. Landsat 7 and its trailing satellite Earth Observing-1 (EO-1) comprise a notable example of a satellite pair that exhibits a trailing formation (Figure 9).

5. PAYLOADS

In the beginning of the satellite-based remote sensing era, the acquired data had a low resolution, and images were blurry compared with aerial images captured by airborne platforms. This situation was due to the limitation of payloads to acquire high-resolution image data from a high altitude, given the perturbations in the atmosphere, such as clouds. However, as payload technology advances, remote sensing satellites have become capable of collecting accurate and high-resolution data not only in black and white or in the visible light region but also in other portions of the EM spectrum beyond what the human eyes can see (Figure 10).

Figure 9. Landsat 7 and EO-1 covering the same ground track at different time intervals (Potter et al., 1998).

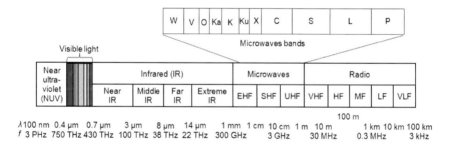

Figure 10. EM spectra commonly used for remote sensing, where λ is the wavelength and f is the frequency of the EM bands in hertz (Hz) (Sellers et al., 2003).

As mentioned earlier, payloads are onboard sensors that detect and collect EM radiation spectra before the received radiation is processed into useable information. However, not all the EM spectra shown in Figure 10 can pass through the atmospheric layer and reach satellites, because some EM bands are blocked by water molecules in the clouds, air molecules and the ozone layer, which are present in the Earth's atmosphere (Sellers et al.,

2003). Therefore, during the payload design process, the EM wavelengths that are blocked and allowed to pass by the atmosphere must be identified such that the reflected and emitted radiation can be detected by sensors. Figure 11 shows the atmospheric windows that indicate the percentage of each EM radiation that passes through the atmosphere.

Atmospheric windows are the windows of transmission or wavelengths of EM bands that pass through the atmosphere by 80% – 100%. As shown in Figure 11, near ultraviolet (NUV), visible light, infrared (IR) and radio waves are the wavelengths that can be used to generate images of the Earth's surface from space. To sense these wavelengths, various types of instruments or sensor systems (Figure 12) have been designed for spaceborne systems.

In general, a satellite is equipped with more than one type of payload system to acquire image data in various EM bands. These systems can be classified into passive or active sensors depending on their operation principle (Baumann, 2010).

5.1. Passive Sensors

Passive sensors require external sources of EM signals, typically the sun, to acquire image data. They rely on reflected or emitted radiation from the targets and work efficiently during daytime, given that the Earth is illuminated by sunlight. These sensors are commonly considered for spaceborne systems, because they require less power supply and are relatively simple to operate. Space cameras and imaging scanners are examples of passive sensors.

5.1.1. Panchromatic Camera

A panchromatic camera is an instrument that images data in grey scale. This type of imagery is useful in geospatial mapping, because the images captured generally have a considerably higher spatial resolution than the multispectral images taken at the same altitude. Consequently, panchromatic cameras are used in modern satellites, such as QuickBird II (resolution: 0.65

m) and IKONOS (resolution: 0.82 m), which capture the Earth's surface in the form of digital panchromatic images for urban and rural mappings (Dial et al., 2003).

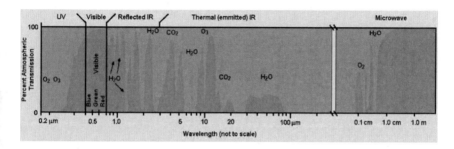

Figure 11. Atmospheric windows that depict some EM bands that readily pass through the atmosphere, whereas some bands are blocked by water and air molecules present in the Earth's atmosphere (Baumann, 2010).

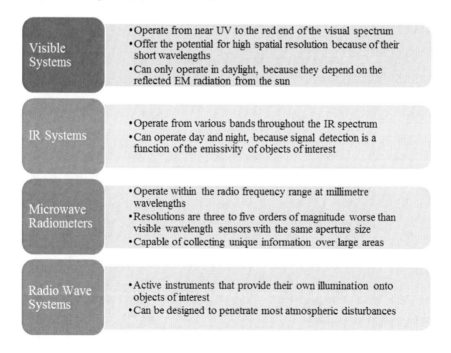

Figure 12. Four types of payload system commonly considered for remote sensing satellites.

5.1.2. Multispectral and Hyperspectral Scanners

Multispectral sensors capture images at a specific frequency of an EM spectrum ranging from visible light to IR bands, typically 10 of the EM bands. Different spectral images of the same object of interest can be generated by separating certain wavelengths with filters or using instruments that are sensitive to specific wavelengths. This technique allows the extraction of additional information, which the human eye cannot capture.

Hyperspectral scanners consist of narrow-band channels (10 – 20 nm) with hundreds of thousands of bands; most of these scanners operate over the visible–near IR/short wave IR (VNIR/SWIR) bands (Shaw et al., 2003). The narrow bands allow a high level of spectral details in the images, thereby making these scanners suitable for finding and differentiating objects that cannot be achieved using multispectral sensors. The common wavelengths of the spectral bands for both scanners with their respective applications are listed in Table 1.

Table 1. Common wavelengths of spectral bands and their uses (Shaw et al., 2003)

Band	Wavelength	Applications
Blue	450 – 515 nm	Atmosphere and deep-water imaging
Green	530 – 590 nm	Vegetation and deep-water imaging
Red	640 – 670 nm	Man-made object imaging
NIR	750 – 900 nm	Primarily for vegetation imaging
Mid IR (MIR)	1550 – 1750 nm	Vegetation, soil moisture content and forest fire imaging
Far IR (FIR)	2080 – 2350 nm	Imaging soil, moisture, geological features and fires
Thermal IR	10400 – 12500 nm	Uses emitted radiation; useful in geological structure imaging, fire detection and night observation
Combined Spectral Bands		
True colours (red, green and blue)		Plain colour photographs, analysis of man-made objects
Green–red–IR		Used in vegetation imaging
Blue–NIR–MIR		Water depth imaging, vegetation coverage and fire detection

Figure 13. True colour image with 40 cm resolution of an airport in Madrid, Spain captured by WorldView-3 (DigitalGlobe, 2013).

WorldView-3 is an example of a high-resolution imaging commercial satellite that uses a multispectral sensor. The payload can detect eight VNIR colours, i.e., red, green, blue, coastal, yellow, red edge, NIR1 and NIR2 bands, and eight SWIR bands that can penetrate haze, fog, smog, dust and smoke, plus twelve CAVIS (Clouds, Aerosols, Water Vapour, Ice and Snow) bands used to map clouds, ice and snow and to correct aerosol and water vapour. Figure 13 shows an image captured by WorldView-3.

5.2. Active Sensors

Active sensors are instruments that can transmit their own EM signals. In general, they have a transmitter unit that transmits EM signals and a receiver unit that collects the reflected or backscattered EM signals from targets. These instruments allow a satellite to acquire image data in the absence of external EM sources. High power supply is required, because the instruments must generate their own EM signals. Radar, synthetic-aperture radar (SAR) and LIDAR are active sensors usually considered for spaceborne geospatial mapping missions.

5.2.1. Radar

Radar (from the acronym **RA**dio **D**etection **A**nd **R**anging) is an object-detection system that uses radio waves or microwaves to determine the range, angle or velocity of an object. This system has a transmitter that transmits radio waves, called radar signals, at a certain frequency that depends on the mission, and a receiver that receives waves, which are reflected off from the target. Radar signals are reflected well by materials with good electrical conductivity, particularly by most metals, seawater and wet ground. In addition, radar signals can detect objects at relatively long ranges due to the weak signal absorption of the medium that it is passing through, in which other EM spectra are strongly attenuated. This property makes radar signals transparent to most atmospheric disturbances, such as clouds, fog, snow and rain (Stimson, 1998). Therefore, radar is suitable for terrain mapping. Commonly used wavelengths in radar imaging are the Ka, X and L bands.

5.2.2. SAR

SAR is a type of radar used to acquire images, such as those of terrain, landscapes and oceans with ground resolutions higher than 5 m, even though a layer of atmospheric disturbances, such as clouds (Johan et al., 1991). In general, SAR produces 2D images; the first dimension is called range (or cross track), which measures the line-of-sight distance from the radar to a target, and the second dimension is azimuth (or along track), which is perpendicular to the range (Figure 14).

SAR can produce a relatively fine azimuth resolution, which basically requires a physically large antenna to focus the transmitted and received radar signals. This requirement is not practical for spaceborne systems. However, SAR overcomes this problem through a technique called coherent processing of the reflected radar signals collected over a long distance along its ground track. This technique results in an increase in antenna size, thereby effectively synthesising a virtual antenna with the same size as the actual physical antenna, which results in a tremendous increase in azimuth resolution (Fitch, 2012). Typical frequency bands used for SAR are listed in Table 2.

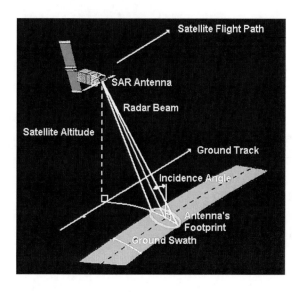

Figure 14. Bandwidth geometry of SAR (Chin, 2001).

Table 2. Frequency bands for SAR (Chin, 2001)

Band	Frequency	Wavelength
P-Band	0.4 – 1 GHz	30 – 75 cm
L-Band	1 – 2 GHz	15 – 30 cm
S-Band	2 – 4 GHz	7.5 – 15 cm
C-Band	4 – 8 GHz	3.7 – 7.5 cm
X-Band	8 – 12 GHz	2.5 – 3.7 cm

Several SARs have flown onboard satellite platforms, such as RADARSAT, which uses a C-band signal, Seasat with an L-band signal and the TerraSAR satellite with an X-band SAR system. Meanwhile, the Korea Multipurpose Satellite-5 (KOMPSAT-5)/Arirang-5 (Figure 15) is the first Korean SAR satellite capable of transmitting X-band signals. The SAR system supports geographic information systems and environment and ocean monitoring with three operation modes, namely high-resolution, standard and wide swath, with 1, 3 and 20 m resolutions, respectively (Lee, 2010). Figure 16 shows one of the SAR images acquired by KOMPSAT-5/Arirang-5.

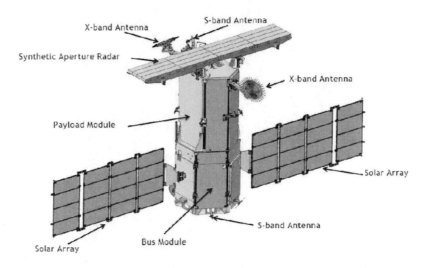

Figure 15. KOMPSAT-5/Arirang-5 with an X-band SAR payload (Lee, 2010; Yoon et al., 2011).

Figure 16. Standard mode SAR image of Sydney, Australia observed in 2013 by KOMPSAT-5/Arirang-5 (Lee, 2010; Yoon et al., 2011).

5.2.3. LIDAR

The LIDAR (from the acronym **L**ight **D**etection **A**nd **R**anging) system uses a narrow laser beam in UV band radiation, visible light or NIR light to image objects of interest in very high resolution. This system can target a wide range of materials, including nonmetallic objects, rocks, chemical compounds, aerosols and cloud, thereby making it suitable for land mapping and atmospheric and meteorological studies. In a LIDAR system, a transmitter unit generates light pulses with wavelengths of a few to several hundred nanoseconds. A beam expander within the transmitter unit is typically applied to reduce the divergence of a light beam before it is sent into the atmosphere. A receiver unit consists of a telescope that collects the backscattered photons from the atmosphere. An optical analysis system selects specific wavelengths or polarisation states from the collected light, depending on the application. The radiation with the selected wavelengths is directed onto a detector, usually a solid-state photodetector, and is converted into an electrical signal to be stored in a computer (Wandinger, 2005). In general, a LIDAR system onboard a satellite requires instruments, such as a Global Positioning System receiver and an inertial measurement unit, to determine the absolute position and orientation of the system. Figure 17 shows the basic setup of a LIDAR system.

An example of a spaceborne LIDAR is ICESat, which is part of NASA's observation system. Its mission is to measure ice sheet mass balance, cloud and aerosol heights and land topography and vegetation characteristics. This satellite is equipped with a sole instrument, called the Geoscience Laser Altimeter System (GLAS), which is a space-based LIDAR combined with a precision surface LIDAR and a sensitive dual-wavelength cloud and aerosol LIDAR (Figure 18).

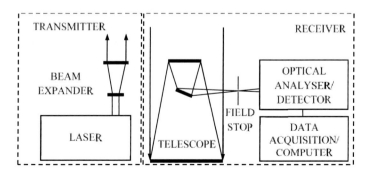

Figure 17. Principle setup of a LIDAR system (Wandinger, 2005).

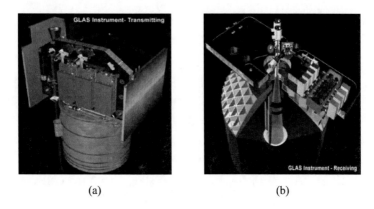

(a) (b)

Figure 18. GLAS instruments mounted onboard ICESat whilst (a) transmitting and (b) receiving (Potter et al., 1998).

GLAS lasers emit IR and visible laser pulses at 1064 nm and 532 nm wavelengths and produce a series of approximately 70 m-diameter laser spots that are separated by nearly 170 m along the satellite's ground track. During its mission, the satellite provides topography data from around the globe, such as mapping of the canopy height of the Amazon rainforest (Sawada et al., 2015) and polar-specific coverage over the Greenland and Antarctic ice sheets.

6. DATA ACQUISITION TECHNIQUES

In general, a remote sensing satellite has its payloads pointing directly towards nadir, i.e., the Earth's centre. Therefore, the data acquired are basically from the region along the satellite's ground track bounded within the field-of-view (FOV) at a certain swath width. Depending on the sensor's design and the satellite's altitude, swath width can vary for each satellite. For the same sensor, satellites with a high-altitude orbit will have a larger swath width compared with a low-altitude satellite, thereby allowing a larger area of coverage for scanning and imaging. In addition, the pointing direction of a sensor can be rotated such that it can make observations at a certain ± angle from the nadir direction. This technique increases the pointing or scanning coverage, which allows a large return of image data, as illustrated in Figure 19. In LEO, remote sensing satellites move with respect to the Earth's surface. This movement along the ground track acts as one of the scan dimensions; therefore, the payload system in this orbit basically needs to perform only 1D scanning, which is in the cross-track direction or range direction within the swath width.

Several scanning techniques can be generally considered to acquire image data, as shown in Figure 20 (Larson et al., 1992). For example, the whisk broom sensors shown in Figure 20(a) scan a single detector element that corresponds to a single pixel on the ground in the cross-track direction. Several detectors can also be used for whisk broom sensors to reduce the requirements compared with those for a single detector. This technique is called multi-element whisk broom sensors (Figure 20(b)). Meanwhile, push broom sensors use a linear arrangement of detectors, called a line of imager, which covers the full swath width (Figure 20(c)). This line of imager moves along the ground track, as satellites pass over the scenes. Step-and-stare sensors use a matrix arrangement of detector elements that covers a part of an image or an entire image within the swath width (Figure 20(d)). The arrangement of detector elements is in a 2D matrix that corresponds to the ground pixel arrangement.

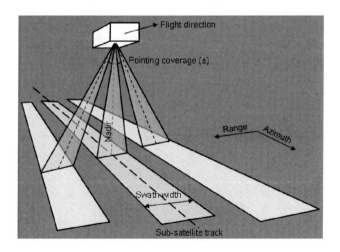

Figure 19. On/off nadir-pointing payload system increases the area of coverage by ± degrees from the nadir.

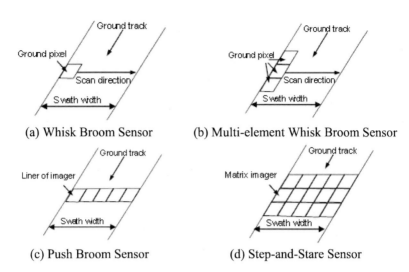

Figure 20. Scanning techniques used to acquire image data for passive payloads, such as a multispectral imager (Larson et al., 1992).

Figure 21. Specially equipped aircraft used to recover film capsules during parachute descent (Baumann, 2009).

7. DATA RECOVERY AND GROUND STATION

In the beginning of the spaceborne remote sensing era, image data were stored in film canisters. The acquired data were retrieved by releasing the film canisters in capsules called 'buckets', which were recovered in mid-air by an aircraft during parachute descent, as shown in Figure 21.

At present, image data are processed and stored onboard the satellite in digital form before being sent back to ground stations on Earth. In general, a satellite is integrated with an onboard communication system, which consists of transmitter and receiver antennas. The stored image data are sent to ground stations in the form of EM signals, typically in the X-band radio wave signals, by the transmitter antenna.

For a LEO satellite, image data can only be downlinked when the satellite is flying above a ground station, because its position in the sky is not fixed with respect to the Earth's surface. To increase the frequency of data retrieval, more than one ground stations are commonly designated to receive the signal. These ground stations are located at different places, and those on the ocean are called ocean-based ground stations. Some satellites, known as relay satellites, transfer signals to other satellites before the signals are relayed to ground stations (Larson et al., 1992). This method increases communication windows and allows image data to be sent in real time. For example, communication satellites in GEO can be used as relay satellites to continuously send image data to a single ground station, thereby reducing the need for multiple ground stations and complex tracking antennas.

Ground stations play an important role in retrieving image data. Moreover, this ground segment element of spaceborne remote sensing systems, called the mission control centre (MCC), is a place where satellite operators manage and operate the mission. Satellite operators work round the clock, planning and scheduling missions, housekeeping, monitoring satellite health, troubleshooting and distributing data to end users.

8. CASE STUDY: LANDSAT 8

Landsat 8 is the eighth and the latest satellite of the Landsat program for Earth observation missions. This satellite was successfully launched on February 11, 2013 by the Atlas V 401 AV-035 rocket from Space Launch Complex 3E, Vandenberg Air Force Base and is currently operating in SSO.

This satellite, shown in Figure 22, was built by the Orbital Sciences Corporation, which served as the prime contractor for the mission, whereas the satellite's instruments were constructed by Ball Aerospace and NASA's Goddard Space Flight Centre (GSFC).

Figure 22. Rendered image of Landsat 8 in SSO (Potter et al., 1998).

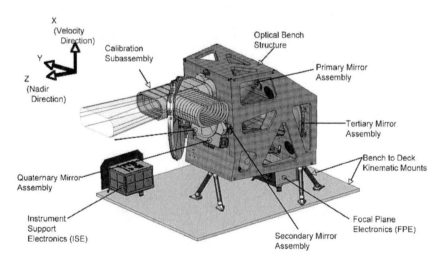

Figure 23. Instrument overview of OLI built by Ball Aerospace (Knight et al., 2014).

Landsat 8's mission is to provide timely, high-quality visible and IR images of all landmass and near-coastal areas on Earth; it continually refreshes and updates an existing Landsat database (Zanter, 2016). The satellite has a design life of 5 years and carries 10 years of consumable fuel supplies. During its mission period, Landsat 8 will provide data continuity for its predecessors (Landsat 4, 5 and 7), offer a 16-day repetitive Earth coverage and periodically refresh a global archive of sun-lit, substantially cloud-free land images.

Two sensors, namely, the Operational Land Imager (OLI) and the Thermal InfraRed Sensor (TIRS) (shown in Figures 23 and 24, respectively), are used onboard Landsat 8 to provide more than 700 scenes daily. The OLI sensor can collect data for nine shortwave spectral bands over a 190 km swath width with a 30 m spatial resolution for all bands, except for the 15 m panchromatic band. This sensor offers improvements over earlier Landsat sensors by using a push broom sensor instead of a whisk broom sensor with a four-mirror anastigmatic telescope, which provides a 15° FOV and covers a 190 km ground swath and 12 bit quantisation, thereby increasing sensitivity, reducing moving parts and improving land surface information (Loveland et al., 2016).

Table 3. Spectral Bands of OLI and TIRS (shown in *italic*) (Zanter, 2016)

Band	Wavelength (μm)	Resolution (m)
Band 1: Coastal/Aerosol	0.435 – 0.451	30
Band 2: Blue	0.452 – 0.512	30
Band 3: Green	0.533 – 0.590	30
Band 4: Red	0.636 – 0.673	30
Band 5: NIR	0.851 – 0.879	30
Band 6: SWIR-1	1.566 – 1.651	30
Band 7: SWIR-2	2.107 – 2.294	30
Band 8: Panchromatic	0.503 – 0.676	15
Band 9: Cirrus	1.363 – 1.384	30
Band 10: TIR-1	*10.60 – 11.19*	*100*
Band 11: TIR-2	*11.50 – 12.51*	*100*

Similar to OLI, TIRS is a push broom sensor that uses quantum well IR photodetectors, which are sensitive to two longwave thermal IR (TIR) bands. The sensor, which has a three-year design life, can collect two TIR bands with a 100 m spatial resolution over a 190 km swath width, thereby allowing separation of the Earth's surface temperature from that of the atmosphere. Table 3 lists the spectral bands of OLI and TIRS.

Image data acquired by Landsat 8 are stored in a solid-state data recorder. Landsat 8 uses an X-band antenna to transmit OLI and TIRS data to its ground station either in real time or in playback mode. The ground system is developed by the United States Geological Survey, which is responsible for monitoring and operating the satellite. Its components include the mission operations element (MOE), collection activity planning element, ground network element and data processing and archive system (DPAS). MOE provides capabilities for command and control, mission planning and scheduling and flight dynamics analysis.

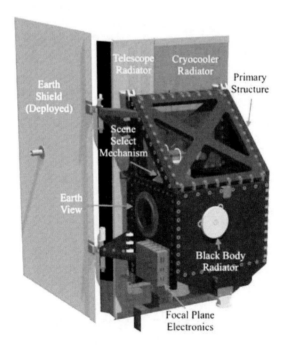

Figure 24. TIRS built by NASA's GSFC, which has a three-year design life (Potter et al., 1998).

Figure 25. Strip images of parts of the western Los Angeles area from the agricultural land near Oxnard captured by Landsat 8 (Clark et al., 1993).

Table 4. Parameters of Landsat 8 (Clark et al., 1993)

Spacecraft Properties	
Manufacturer	Orbital Sciences (prime) Ball Aerospace (OLI) NASA GSFC (TIRS)
Launch mass	2623 kg
Dry mass	1512 kg
Launched date	11 February 2013, 18:02 UTC
Entered service	30 May 2013
Orbital Parameters	
Reference system	Geocentric
Regime	SSO
Nominal altitude	705 km
Semi-major axis	7080.48 km
Eccentricity	0.0001310
Inclination	98.2248°
Ascending node (RAAN)	219.1493°
Argument of perigee	91.0178°
Mean anomaly	269.1170°
Period	98.8 minutes

Meanwhile, DPAS is responsible for ingesting, archiving, calibrating, processing and distributing Landsat 8 data. It also has a direct portal to user communities. Figure 25 shows images of the Earth captured by Landsat 8, whereas the parameters of Landsat 8 are listed in Table 4. This strip of images is part of the western Los Angeles area, from the agricultural land near Oxnard. The image on the left is in true visible RGB colours (Bands 2, 3 and 4). It shows most of the urban area in light grey. The image in the middle is in NIR Band 5, in which bright features denote parks and heavily irrigated vegetation. The image on the right is in SWIR-1 and SWIR-2 (Bands 6 and 7). It shows vegetation activities in bright green spots.

Acknowledgment

The authors acknowledge the support provided by Universiti Sains Malaysia's Bridging Grant 304/PAERO/6316112 in the execution of this work.

References

Baumann. (2009). *History of remote sensing, satellite imagery*, part II. Last modified.

Baumann. (2010). *Introduction to Remote Sensing*. State University of New York, College at Oneonta.

Chin. (2001). *Synthetic Aperture Radar (SAR)*, from https://crisp.nus.edu.sg/~research/tutorial/mw.htm.

Clark, Swayze., King, Gallagher. & Calvin. (1993). *The US Geological Survey, digital spectral reflectance library*: Version 1: 0.2 to 3.0 microns.

DigitalGlobe. (2013). Basic Imagery.

Fitch. (2012). *Synthetic aperture radar*: Springer Science & Business Media.

Johan. & Robert. (1991). *Synthetic aperture radar*: John Wiley & Sons, Inc, New York.

Knight. & Kvaran. (2014). Landsat-8 operational land imager design, characterization and performance. *Remote Sensing*, 6(11), 10286-10305.

Kramer. (2002). *Observation of the Earth and its Environment: Survey of Missions and Sensors*: Springer Science & Business Media.

Larson, & Wertz. (1992). *Space mission analysis and design*: Torrance, CA (United States); Microcosm, Inc.

Lee. (2010). Overview of KOMPSAT-5 program, mission, and system. Paper presented at the *2010 IEEE International Geoscience and Remote Sensing Symposium*.

Loveland. & Irons. (2016). Landsat 8: The plans, the reality, and the legacy. *Remote Sensing of Environment*, *185*, 1-6.

Panissal, Brossais. & Vieu. (2010). Les nanotechnologies au lycée, une ingénierie d'éducation citoyenne des sciences [Nanotechnology in high school, an engineering of citizen science education]. *RDST. Recherches en didactique des sciences et des technologies*, (1), 319-338.

Potter, Gasch. & Bauer. (1998). *A photo album of earth scheduling Landsat 7 mission daily activities*.

Sabol, Burns. & McLaughlin. (2001). Satellite formation flying design and evolution. *Journal of spacecraft and rockets*, *38*(2), 270-278.

Sawada, Suwa., Jindo, Endo., Oki, Sawada., Arai, Shimabukuro Celes. & Campos. (2015). A new 500-m resolution map of canopy height for Amazon forest using spaceborne LiDAR and cloud-free MODIS imagery. *International journal of applied earth observation and geoinformation*, *43*, 92-101.

Sellers, Astore., Giffen, Larson., Sellers, Astore Giffen. & Larson. (2003). *Understanding Space: An Introduction to Astronautics* 2nd Ed.

Shaw. & Burke. (2003). Spectral imaging for remote sensing. *Lincoln laboratory journal*, *14*(1), 3-28.

Stimson. (1998). *Introduction to Airborne Radar* (2nd E-dition): North Carolina: Scitech Publishing, Inc.

STK. (2019). *AGI STK Images*, 2017, from https://www.agi.com/products/engineering-tools.

Wandinger. (2005). Introduction to lidar *Lidar*, (pp. 1-18), Springer.

Yoon, Keum., Shin, Kim., Lee, Bauleo., Farina, Germani Mappini. & Venturini. (2011). Kompsat-5 SAR design and performance. Paper presented at the *2011 3rd International Asia-Pacific Conference on Synthetic Aperture Radar (APSAR)*.

Zanter. (2016). Landsat 8 (L8) data users handbook. *Landsat Science Official Website*. Available online: https://landsat. usgs. gov/landsat-8-l8-data-users-handbook (accessed on 20 January 2018).

ABOUT THE EDITORS

Dr. Parvathy Rajendran
School of Aerospace Engineering, Universiti Sains Malaysia,
Pulau Pinang, Malaysia
Email: paru80_2000@hotmail.com

Dr. Parvathy Rajendran (https://www.researchgate.net/profile/Parvathy_Rajendran2) is an associate professor and has served in the School of Aerospace Engineering at Universiti Sains Malaysia since 2013. She completed her Ph.D. in Aerospace Engineering from Cranfield University, United Kingdom in October 2012. There, her research includes UAV design, development and flight testing and UAV's systems development and testing. She has also been involved as a systems and flight test engineer in various UAV projects in the UK, which all successfully flown. Her appointment as the Head of System and Design Research Cluster and as the Laboratory Manager for UAV Laboratory in the School of Aerospace Engineering provides further recognition to her credibility. In addition, she has attained the titles Professional Engineer from the Board of Engineers Malaysia and Chartered Engineer (CEng.) from the Engineering Council in United Kingdom. As a young research academician, Rajendran has produced many high-impact publications. She has also been appointed by few international journals to be an editor-in chief, guest editor, international advisor and member of the reviewer board. She has been the chairman and member of the technical conference committee of various international conferences. In addition, she has maintained various grants, which totals more than RM 1 million. Thus, she has contributed well for the development of human capitals. Her active involvement in teaching, research, consultancy and service to the university and community has emphasized her enthusiastic academician nature with high potential to contribute on the aerospace engineering field.

Dr. Muhd Zulkifly Abdullah
School of Mechanical Engineering, Universiti Sains Malaysia,
Engineering Campus, Pulau Pinang, Malaysia

Dr. M.Z. Abdullah (https://www.researchgate.net/profile/MZ_Abdullah) has been a professor of mechanical engineering at Universiti Sains Malaysia since 2010. He obtained his B.S. degree in mechanical engineering from the University of Swansea, UK. He took up his MSc and Ph.D. degrees from the University of Strathclyde, UK. He has numerous publications in international journals and conference proceedings. His areas of research are CFD, heat transfer, electronic packaging, electronic cooling, and porous medium combustion. He has successfully established collaborations with Intel Technology (M) Sdn. Bhd., Celestica (M) Sdn. Bhd., and Jabil Circuits (M) Sdn. Bhd. in the areas of electronic packaging and surface mount technology. He is also involved in consultancy work with five companies in Malaysia related to his field of expertise. He has been invited to give keynote lectures at University of Manipal in India, National Institute of Technology in Calicut, India, University of Tarumanangara in Jakarta, Indonesia, University of Kuala Lumpur (MSI) in Kedah, Malaysia, and Intel Technology (M) Sdn. Bhd. in Penang, Malaysia. Ir. Dr. Abdullah has supervised 21 Ph.D. and more than 30 MSc students. Currently, he is supervising seven Ph.D. students in his research laboratory. He was the recipient of the Outstanding Paper Award 2013 conferred by Emerald Group Publishing Limited in the journal of Soldering and Surface Mount Technology.

INDEX

A

abrasion, 130, 132, 133, 136
absorption, 52, 73, 74, 85, 90, 96, 97, 102, 108, 109, 110, 111, 112, 115, 117, 118, 119, 120, 200
actuator, 160, 165
advance ratio, 154, 163
aeroelastic, 159
aero-engine(s), v, 123, 124, 126, 127, 128, 140, 148
aerospace, v, vii, viii, 1, 2, 4, 5, 11, 35, 45, 46, 48, 51, 73, 75, 77, 90, 118, 120, 121, 123, 146, 148, 149, 163, 165, 167, 185, 208, 211, 215
airfoil, vii, 136, 137, 142, 143, 145, 150, 151, 152, 155, 164
airworthiness, 125, 169, 170, 182
antenna, 77, 200, 207, 210
attitude and orbit control system (AOCS), 189
autonomous, 171, 172
aviation fuel, v, 73, 74, 75, 77, 78, 82, 83, 90, 91, 92, 110, 111, 115, 116, 138
aviation jet fuel, 73

B

ball grid array (BGA), 8, 14, 19, 27, 30, 31
biodiesel, 74, 78, 79, 80, 81, 82, 83, 84, 90, 93, 109, 110, 114, 115, 117, 118, 119, 120, 121, 122
blade, 129, 132, 133, 136, 137, 139, 140, 141, 142, 143, 145, 146, 147, 161
blade clearance, 141
blended biodiesel(s), 83
boundary layer, 153, 159
brittle, 64, 91, 128, 135, 147

C

cameras, 171, 172, 173, 174, 196
carbon deposition, 139
carbon fibre, v, 52, 69, 76, 89, 93, 115, 119, 121
carbon nanotubes, 51, 52, 53, 54, 58, 66, 67, 68, 69, 70, 71
casing distortion, 138
centrifugal force, 140
choking, 138
classical Fickian, 87, 88

combustion, 77, 123, 127, 128, 131, 132, 138, 139, 147, 148, 169
combustion chamber liner(s), 131
communication and data handling system (CDHS), 189
composite material(s), 51, 52, 73, 74, 75, 76, 87, 91, 99, 102, 108, 109, 110
compressor washing, 124, 147
corrosive, 93, 107, 112
corrugated wing, 155
creep, 129, 133, 135, 136, 140, 146

D

damage tolerance, 51, 52
data acquisition, 185, 187, 204
degradation of mechanical properties, 74
deposition, 53, 66, 124, 125, 137, 139, 146
design of experiment (DOE), 11, 17, 19
deterioration, v, 123, 124, 125, 128, 130, 132, 136, 137, 138, 139, 140, 141, 142, 143, 144, 146, 147, 148
diffusion, 74, 84, 85, 86, 87, 88, 90, 93, 99, 106, 107, 108, 109, 111, 112, 118, 119, 120, 121
downstroke, 151, 152, 154, 156
drone, 168, 172, 180

E

efficiency, 57, 75, 76, 77, 80, 123, 126, 127, 136, 137, 139, 141, 142, 144, 153, 160, 161, 162, 173, 178, 182
electromagnetic, vii, 47, 49, 70, 187
electronical power system (EPS), 189
engine health monitoring, 125, 147, 148
engine performance, 80, 84, 121, 124, 125, 127, 130, 142, 144, 146, 147
environment-friendly, 16, 79

erosion, 124, 125, 130, 132, 133, 136, 137, 140, 141, 143, 144, 146, 148
exhaust gas temperature, 123, 126

F

factor(s), 2, 14, 16, 25, 77, 85, 86, 88, 107, 111, 123, 124, 125, 128, 135
fatigue, 17, 25, 51, 63, 91, 113, 118, 121, 129, 133, 134, 135, 140, 145, 146, 147, 148
FE method (FEM), 13, 19
Fickian curve, 85, 90
Fickian model, 87
filtration, 81, 124
finite volume (FLUENT), 2, 13, 14
flapping cycle, 155, 156
flapping frequency, 152, 153
flapping wing(s), 149, 150, 153, 154, 156, 157, 158, 159, 160, 161, 162, 164, 165
flexural strength, 59, 97, 98
flexural stress, 91, 92, 101, 102, 140
flight path, 171, 179, 180, 189, 192
flow unsteadiness, 150, 153, 154, 162
fluid–structure interaction (FSI), 2, 14, 15, 16, 19, 26, 32
fluxing, 7, 17
foreign object damage, 124
formation flying, 189, 192, 194, 213
fracture toughness, 52, 56, 60, 61, 62, 63, 64, 69, 93, 134
friction, 128
FSI simulation, 2, 14, 15
fuel attack, 74, 108
fuel consumption, 123, 127, 142
fuel uptake, 73, 74, 108

Index 221

G

geographic information system (GIS), 177
georeferencing, 176
geospatial mapping, 167, 168, 169, 170, 172, 174, 176, 185, 186, 187, 188, 196, 199
geostationary earth orbit (GEO), 190, 207
GFRP composites, 97, 102
glass fibre epoxy laminates, 91
glass transition temperature, 93, 94, 115
global navigation satellite system (GNSS), 179
GPS, 172, 173, 177, 178, 179, 180
grazing, 132
ground radio controller, 172, 179
ground stations, 187, 207

H

heat stress, 123
hierarchical composite(s), 51, 52, 54
hot–wet, 74

I

immersion, 89, 90, 91, 92, 94, 95, 96, 97, 99, 100, 102, 103, 104, 105, 106, 107, 108, 110, 113, 114, 115, 116
impinging, 132, 133
infrared reflow, 20
instabilities, 123, 138
Interconnection, v, 1
inter-laminar shear strength, 94

K

kerosene, 78, 80, 82, 108, 110, 113, 117, 119, 121
knitted composite, 35, 46, 47
knitted structure(s), v, 35, 38, 39, 42, 45, 47, 48

L

Landsat 8, 187, 207, 208, 209, 210, 211, 213, 214
leading-edge vortex (LEV), 153, 158, 166
lifespan, 123
light detection and ranging (LIDAR), 199, 202, 203
low earth orbit (LEO), 190, 192, 205, 207

M

machine vision, 168
maintenance, 124, 146
mechanical properties, 32, 46, 47, 48, 50, 52, 53, 54, 56, 57, 60, 66, 69, 74, 87, 91, 93, 97, 102, 107, 108, 109, 110, 112, 115, 116, 117
mechanical property degradation, 74
mechanism(s), 7, 23, 24, 57, 60, 62, 63, 64, 66, 67, 70, 91, 111, 119, 124, 129, 130, 131, 132, 136, 137, 141, 142, 146, 147, 154, 157, 158, 159, 162, 164, 166, 189
microelectronics, 1, 4, 13, 27, 28, 29, 31
mission, vii, 128, 144, 167, 168, 170, 171, 172, 173, 174, 175, 176, 177, 178, 179, 180, 187, 188, 189, 192, 200, 203, 204, 207, 208, 209, 210, 213

moisture diffusion, 73, 74, 84
molten solder, 2, 7, 12, 16, 26
morphing, 153
multiaxial fabric, 35

N

nanocomposite(s), vii, 60, 97, 111
non-Fickian sorption curve, 85, 87, 88
non-recoverable, 124

O

oblique photograph, 175
off-wing, 124
on-wing, 124
orbit, 185, 186, 187, 188, 189, 190, 191, 192, 193, 204
orbital, 183, 190, 192, 193, 208, 211

P

panchromatic camera, 196
passive sensor, 196
path planning, 176, 179, 180, 182
payload, 170, 171, 172, 173, 175, 184, 188, 189, 194, 196, 197, 199, 202, 205
PCB assembly, 4
PCB hole, 2, 7, 11, 16
PCB warpage, 17
performance, v, 1, 2, 3, 4, 11, 16, 17, 19, 27, 30, 31, 36, 51, 52, 54, 56, 62, 66, 67, 74, 75, 76, 77, 84, 90, 109, 110, 113, 114, 115, 123, 124, 125, 126, 127, 130, 136, 137, 139, 141, 143, 145, 146, 147, 148, 155, 160, 161, 162, 165, 170, 172, 173, 174, 213, 214

permanent, 4, 87, 94, 99, 108, 124, 130, 138
petroleum, 74, 78, 80, 81, 82, 112, 118, 120
pin-through hole (PTH), 1, 3, 4, 5, 7, 11, 12, 13, 26, 28, 31
pitching, 155, 156, 157, 165
plasticiser effect, 94, 98, 106
polar orbit, 190, 192
polymer composites, 51, 54, 62, 68, 116, 119, 121
power, 4, 5, 19, 73, 123, 146, 147, 161, 171, 188, 189, 191, 196, 199
preheating, 5, 7, 11
pressure gradient, 155, 159
PTH connector, 2, 11
PTH solder joint, 2, 4, 11, 31

Q

quasi-steady, 150, 160, 161

R

radar, 199, 200, 212
recoverable, 124, 130
regulations, 124, 169, 170, 172
remote sensing, 150, 170, 181, 185, 186, 187, 188, 189, 190, 193, 194, 195, 204, 206, 207, 212, 213
Reynolds number, 150, 153, 164, 165
Rotating wings, 153
rub strip erosion, 137

S

satellite constellation, 192, 193
satellites, vii, 77, 186, 187, 189, 190, 191, 192, 193, 194, 195, 196, 197, 204, 205, 207

Index

saturation phase, 85
scanners, 198
sensor, 172, 174, 196, 199, 204, 209, 210
solder joint, 1, 2, 3, 4, 5, 6, 7, 11, 12, 16, 17, 18, 19, 20, 23, 26, 32, 34
solder joint defect(s), 7, 12, 20, 23
solder joint failure, 17
soldering machine, 2, 3, 7, 20
space technology, 185
spaceborne, 185, 187, 196, 199, 200, 203, 206, 207, 213
spacer fabric, 35, 47, 48
spray pattern, 138, 140
sulphidation, 131
sun-synchronoous orbit (SSO), 190, 191, 207, 208, 211
surface finish degradation, 125
surface mount technology (SMT), 4, 5, 6, 31

T

take-off incursion, 129
temperature, 2, 5, 7, 9, 12, 13, 16, 17, 19, 21, 22, 23, 24, 26, 29, 32, 55, 77, 78, 79, 80, 81, 86, 93, 94, 108, 109, 116, 119, 121, 123, 125, 127, 128, 129, 130, 131, 132, 135, 138, 139, 140, 141, 143, 144, 146, 210
textile fabrics, 35, 36
THC, 5, 7
thermal distress, 123, 130, 131, 132, 136, 138
thermal expansion, 11, 17, 19, 25
thermoset composite, 21, 22, 30
through-hole (THT), 3, 4, 5, 6
tolerance, 52, 53, 129, 178
topography, 186, 203, 204

trailing formation, 192, 194
turbine disks, 131
turbine vanes, 131, 142

U

unmanned aerial vehicles (UAV), vii, 149, 150, 154, 168, 169, 170, 171, 172, 173, 174, 175, 176, 178, 179, 180, 181, 182, 183, 184, 187
unsteady aerodynamic forces, 151, 158
unsteady aerodynamics, 150, 161
upstroke, 152, 154, 156, 157

V

vane twisting, 141
vertical fill, 2, 11
viscous, 55, 85, 93, 153, 159
visual localisation, vi, 167
vortex, 125, 153, 154, 158, 166

W

wake, 153, 154, 161
warp knitting, 35, 36, 41, 42, 44, 47
water environment, 94, 98, 99, 113
wave soldering, 2, 3, 4, 5, 7, 8, 9, 10, 11, 12, 13, 14, 16, 17, 19, 20, 21, 22, 23, 24, 25, 27, 28, 31, 32, 33
weft knitting, 35, 36, 37, 38, 41, 42, 43
wing kinematics, 153, 156, 158, 161
wing rotation, 159
wing twist, 159

Z

zero-defect, 3

Related Nova Publications

THE INTERNATIONAL SPACE STATION: MANAGEMENT AND UTILIZATION ISSUES FOR NASA

EDITOR: Evelyn Clemens

SERIES: Space Science, Exploration and Policies

BOOK DESCRIPTION: This book assesses the extent to which CASIS has implemented the required management activities; and NASA and CASIS measure and assess CASIS's performance.

HARDCOVER ISBN: 978-1-63484-071-2
RETAIL PRICE: $140

EXOPLANETS – EXTANT LIFE?

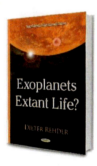

AUTHOR: Dieter Rehder

SERIES: Space Science, Exploration and Policies

BOOK DESCRIPTION: The present book addresses the formation of planetary systems in the wake of collapsing interstellar gas and dust clouds, and the generation as well as the survival and germination of simple molecules serving as modules for more complex molecular constructs that constitute life.

HARDCOVER ISBN: 978-1-63463-301-7
RETAIL PRICE: $110

To see complete list of Nova publications, please visit our website at www.novapublishers.com

Related Nova Publications

INTEGRATION OF CIVIL UNMANNED AIRCRAFT SYSTEMS INTO THE NATIONAL AIRSPACE SYSTEM: ROADMAP, PLANS AND PRIVACY

EDITOR: Jessica Rivera

SERIES: U.S. Transit, Transportation and Infrastructure: Considerations and Developments

BOOK DESCRIPTION: This book focuses on the integration of civil unmanned aircraft systems in the national airspace system; provides an unmanned aircraft systems comprehensive plan; and discusses the final privacy requirements for the unmanned aircraft system test site program.

HARDCOVER ISBN: 978-1-62948-996-4
RETAIL PRICE: $130

ADVANCES IN AEROSPACE SCIENCE AND TECHNOLOGY

EDITOR: Parvathy Rajendran and M.Z. Abdullah

SERIES: Mechanical Engineering Theory and Applications

BOOK DESCRIPTION: In this book, the most accurate and current materials were gathered, reviewed, and presented by an exceptional group of experts. This book presents state-of-the-art, current developments and applications in aerospace.

HARDCOVER ISBN: 978-1-53611-099-9
RETAIL PRICE: $195

To see complete list of Nova publications, please visit our website at www.novapublishers.com